Pr

THE WELFARE OF PIGS

CURRENT TOPICS IN VETERINARY MEDICINE AND ANIMAL SCIENCE

VOLUME 11

Other titles in this series

Volume 1
Control of Reproduction in the Cow
edited by J.M. Sreenan

Volume 2
Patterns of Growth and Development in Cattle
edited by H. de Boer and J. Martin

Volume 3
Respiratory Diseases in Cattle
edited by W.B. Martin

Volume 4
Calving Problems and Early Viability of the Calf
edited by B. Hoffmann, I.L. Mason and J. Schmidt

Volume 5
The Future of Beef Production in the European Community
edited by J.C. Bowman and P. Susmel

Volume 6
Diseases of Cattle in the Tropics: Economic and Zoonotic Relevance
edited by M. Ristic

Volume 7
Control of Reproductive Functions in Domestic Animals
edited by W. Jöchle and D.R. Lamond

Volume 8
The Laying Hen and its Environment
edited by R. Moss

Volume 9
Epidemiology and Control of Nematodiasis in Cattle

Volume 10
The Problem of Dark-Cutting in Beef
edited by D.E. Hood and P.V. Tarrant

Series ISBN: 90-247-2429-5

THE WELFARE OF PIGS

A Seminar in the EEC Programme of
Coordination of Research on Animal Welfare
held in Brussels, November 25-26, 1980

Sponsored by the Commission of the
European Communities, Directorate-General
for Agriculture, Coordination of Agricultural
Research

edited by

W. SYBESMA
Research Institute for Animal Husbandry 'Schoonoord'
Zeist, The Netherlands

1981
MARTINUS NIJHOFF PUBLISHERS
THE HAGUE / BOSTON / LONDON
for
THE COMMISSION OF THE EUROPEAN COMMUNITIES

Distributors

for the United States and Canada
Kluwer Boston, Inc.
190 Old Derby Street
Hingham, MA 02043
USA

for all other countries
Kluwer Academic Publishers Group
Distribution Center
P.O. Box 322
3300 AH Dordrecht
The Netherlands

This volume is listed in the Library of Congress in Cataloging in Publication Data

ISBN 90-247-2521-6 (this volume)
ISBN 90-247-2429-5 (series)

Publication arranged by
Commission of the European Communities,
Directorate-General Scientific and Technical Information and Information Management

EUR 7327 EN

Manuscript Preparation by
Janssen Services, 33a High Street, Chislehurst, Kent BR7 5AE, UK

LEGAL NOTICE

Neither the Commission of the European Communities nor any person acting on behalf of the Commission is responsible for the use which might be made of the following information.

PRINTED IN THE NETHERLANDS

CONTENTS

PREFACE

In Europe, at the present time, there is a growing concern about the welfare of animals kept for production purpose. Technical advancements make it possible to produce more and more efficiently. However, the basic requirements for animal well-being are not always taken into account, partly because they need to be defined more precisely in order to be able to implement them.

In the meeting on the welfare of pigs, organised under the supervision of the Commission of the European Communities, Directorate General for Agriculture, Division for the Coordination of Agricultural Research, scientists from all over Western Europe brought together relevant experimental data of physiological, veterinary, ethological and economic origin for a multidisciplinary discussion on this subject. The results of research studies were also reviewed in the light of possible practical applications and future research.

The meeting was a fruitful first step in bringing together different opinions in order to arrive at a universally acceptable common solution for modern pig production.

W. Sybesma

SESSION I

BEHAVIOURAL AND PHYSIOLOGICAL NEEDS OF PIGS IN GENERAL
I. BASIC NEEDS

Chairman: J.P. Signoret

BEHAVIOURAL AND REPRODUCTIVE STRATEGIES IN WILD FORMS OF *Sus scrofa* (EUROPEAN WILD BOAR AND FERAL PIGS)

R. Mauget

Centre d'Etudes Biologiques des Animaux Sauvages,
79360 Beauvoir, France.

1. INTRODUCTION

The aims of modern pig rearing are directed toward a constant improvement of productivity. The resulting rearing conditions are more and more removed from the original way of life of the wild animal. Thus, a knowledge of the 'possibilities' of the wild species appears to be fundamental, in order to allow the rational development of rearing systems to which the animal can adapt and which will give optimum production.

In contrast to other domesticated species, data are available on the pig's wild ancestor. The aim of this paper is to give an overview of our knowledge of behavioural and physiological mechanisms involved in the adaptation of wild animals to different environmental conditions. For every mechanism we will emphasise the range of spontaneous variation and consequently the possibilities for adaptation.

2. SOCIAL BEHAVIOUR

From a limited number of studies, carried out either in zoological gardens (Frädrich, 1965; Gundlach, 1968) or in a natural environment (Kurz and Marchinton, 1972; Beuerle, 1975; Barrett, 1978; Meinhardt, 1978; Mauget, 1980), it is possible to characterise the social system (Figure 1) showing the group composition, the organisation (hierarchy) and the interindividual bonds.

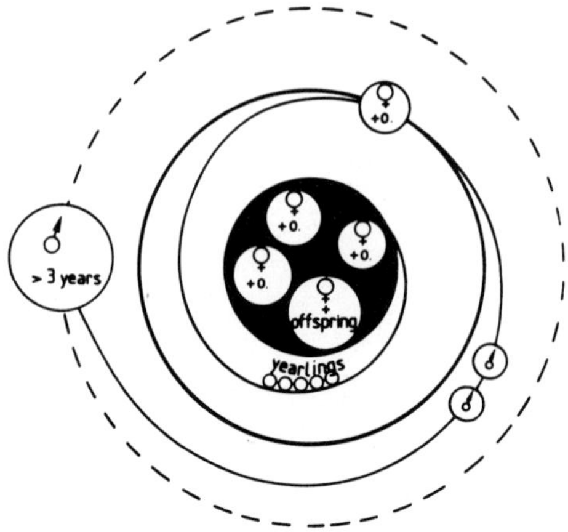

Fig.1. Schematic representation of the structure and organisation of a group. Each individual is represented by a cicle proportional to its social rank. The shaded area represents the nucleus. Peripheral animals are located on a spiral representing the possibility of a temporary association with the central group.

The basic social unit is the female with her offspring of the year. The grouping of a small number (rarely more than 4) of such basic units constitutes the nucleus of a group. These animals live in a close association and appear to have relatively strong interindividual bonds. Their time and space utilisation are synchronous. Near this nucleus 'peripheral animals' are found: some of the young of the previous year and adults (females or males) which are loosely associated with the females of the nucleus. Variations in group composition can occur with respect to the number of females in the nucleus (2 - 4) and the number and sex of the peripheral individuals.

A stable linear hierarchy is revealed by the agonistic interactions which occur mainly in situations where there is competition for food. The most important factors governing social order are sex and, within each sex group, the closely related factors of weight, size and strength. Males over 3 years of age dominate all other classes. The dominant female of the nucleus is usually the oldest. The number of interactions between group members is higher in large groups.

A large variation in social systems among wild ungulates has been described, from the simplest - isolated black rhinoceros females - to the elaborate organisation of wildebeest herds. The social organisation of wild boars and feral pigs appears to be at the lowest end of the scale of sociability and is extremely variable, allowing adaptation to a wide range of environments. Thus, it would seem likely that the small group size is related to the type of habitat. In wooded dense cover habitats, small grouping may represent an adaptive strategy correlated to density of resources and protection against (human) predation. The grouping of several females and their offspring also appears to be beneficial in that it provides mutual protection of the young. Furthermore, reproduction is synchronised within groups. Communal care of the young can be observed, and even cross-suckling between litters. The status of adult males, solitary and nomadic, permits a promiscuous mating strategy, resulting in a continuous mixing in the gene pool of the species.

Several aspects are of importance when comparing these facts with the situation of commercial pig rearing: the female social group is relatively loose, tolerating the integration of new members with ease; the young from several litters are generally reared in a communal group, with the males reaching their solitary social position after several years; batch farrowing is part of the normal reproduction system of wild pigs.

3. SPACE AND TIME UTILISATION

Space and time utilisation are closely related behavioural strategies which are deeply influenced both by environmental factors (biological and physical) and internal (physiological) factors. We will limit our studies to the most important factors involved in the regulation of these behavioural strategies. Most of these data were obtained from radiotelemetry investigations.

3.1. Space utilisation

Habitat selection is strongly influenced by phytostructural characteristics. Thus, bedding sites are located exclusively in heavy thickets or dense copsewood. Conversely, open vegetational structures are preferred for feeding (Figure 2).

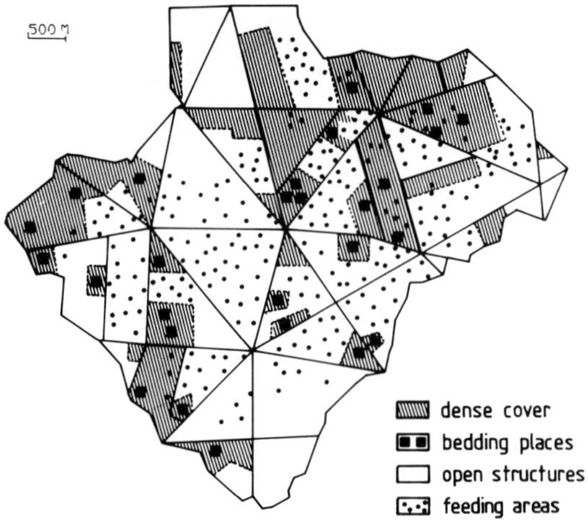

Fig.2. Bedding and feeding areas of radiotracked animals.

Food availability is the main environmental factor modulating the shape, the size and the structure of the home range. It appears from the literature that the home range of sedentary populations of wild boars or feral pigs varies from 100 to 2 500 ha (Kurz and Marchinton, 1972; Pine and Gerdes, 1973; Martin, 1975; Barrett, 1978; Mauget, 1979; Wood and Brenneman, 1980). It is not a fixed characteristic of the species. Such a wide variation may be the result of variations in the resources present in the habitat - the greater the food supply, the smaller is the home range.

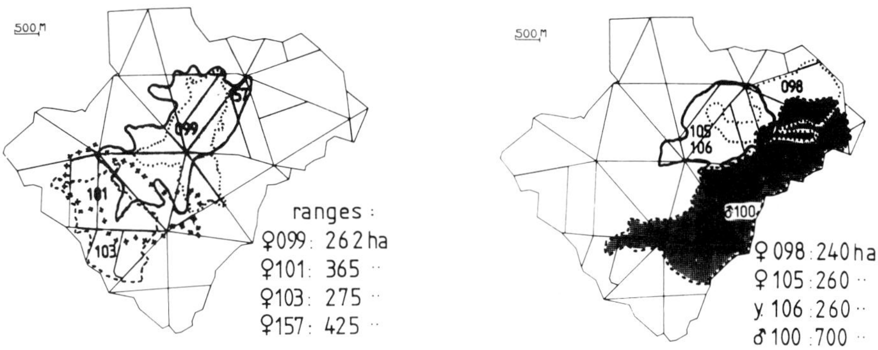

Fig.3. Home ranges of 8 radiotracked wild boars.

Sex differences in the size of the home range have been observed (Figure 3). The size of the home range is larger for boars than for sows. This may be related to the social organisation: the small permanent female groups are relatively sedentary whereas solitary adult males appear to be mostly nomadic.

Farrowing results in profound modifications of the utilisation of space. A drastic reduction (from several hundred ha to 1 ha) occurs in the home range of females in the later stages of pregnancy, simultaneously with a temporary break with the social group. A nest is then built (protective strategy) in this 'farrowing range'. After farrowing, the sow and her litter remain in a limited area, or 'nursing range', of about 20 ha for 1 - 2 weeks.

According to our data, the European wild boar in the Chizé forest exhibits daily nocturnal activity. The activity period (from leaving the sleeping place to re-entering it) each day occurs between sunset and sunrise (Figure 4). A significant correlation between the onset of activity and sunset has been established, but a synchronisation of the cessation of activity with sunrise is less obvious.

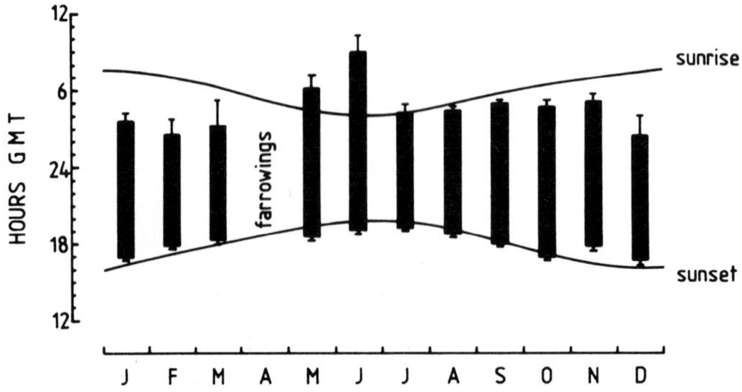

Fig.4. Monthly variations ($\bar{x}$ ± S.E.) of daily activity (pooled data from 6 wild boars.

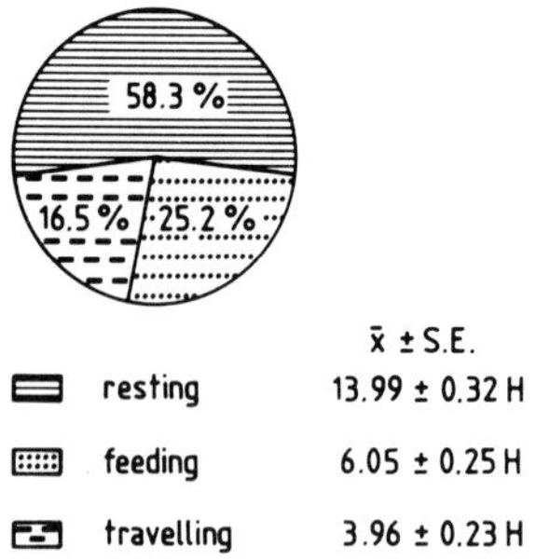

Fig.5. Composite time budget, from 512, 24 h activity patterns.

A great number of suidae species are characterised by a diurnal activity. Such a pattern has been demonstrated in the warthog (Frädrich, 1965; Clough and Hassam, 1970) and in the foresthog (D'Huart, 1976). Feral hogs have been classified as either permanently diurnal (Hanson and Karstad, 1959) or temporarily diurnal, i.e. diurnal during winter but shifting to nocturnal activity during the summer months (Barrett, 1978). According to Briedermann (1971) the nocturnal activity of the wild boar might be atypical, being related to human hunting pressure.

The composite time budget (Figure 5) indicates that, in general, the wild boar spends more time resting than being active (feeding and travelling). This relatively long duration of bedding time may be interpreted as a general tendency of the species to conserve energy. It is worth noting that when food is supplied in large amounts the resting time can increase to up to 80% of the time.

The birth of young induces variations in the utilisation of space. The usual activity rhythm is altered during the nursing period and becomes polyphasic.

The use of space and time appears to be characterised by a high level of adaptability. The home range size is determined by the availability of food and a safe resting place. Protection against low temperature is only required during the first week after farrowing and is provided by the nest built by the female. The search for food appears to be the major factor which controls the time budget.

4. REPRODUCTION

Whereas in modern breeds of domestic swine the litter may contain as many as 20 young, in the wild boar it contains on average 4.6 young (Mauget, 1972). However, intrauterine mortality, which reaches 33% in the domestic sow, is significantly lower (less than 20%). Finally, an essential trait of the wild boar reproduction pattern is that it is seasonal.

Depending on the year, 2 types of farrowing distribution can be observed (Figure 6). In the unimodal type, the peak of farrowing occurs in April - May. In the bimodal type, the first peak occurs earlier, in January - February, and the second in August - September. The animals involved in those 2 main periods are young females reaching puberty at one of those peaks and adult females farrowing twice a year (Mauget, 1978).

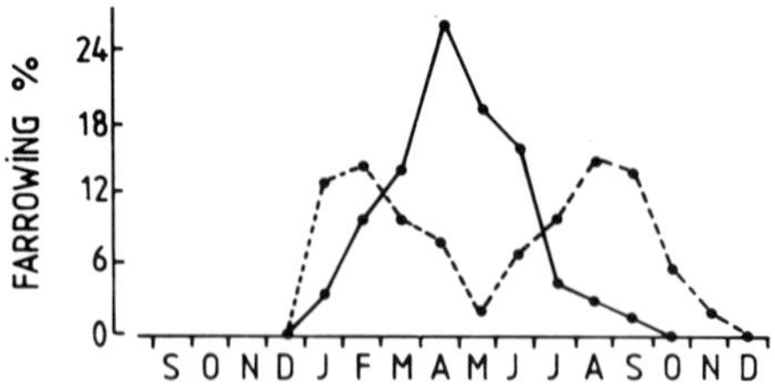

Fig.6. Monthly distribution of farrowings.
(----) unimodal type, data from 1 584 births over 7 years.
(- - -)bimodal type, data from 401 births over 3 years.

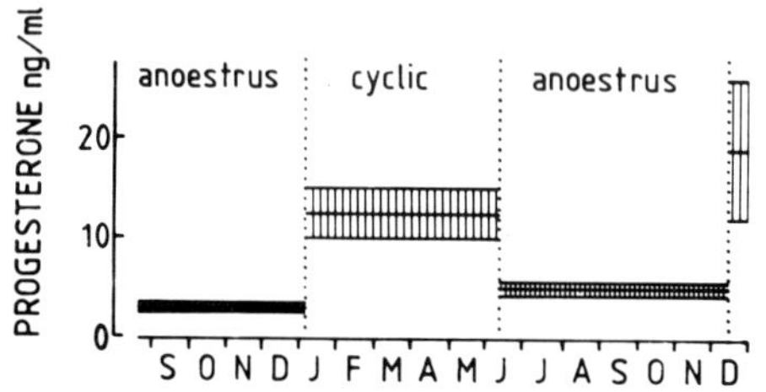

Fig.7. Seasonal variations in ovarian activity as determined by mean progesterone levels ($\bar{x} \pm 2$ S.E. : 0.95 confidence limits).

Longitudinal analysis of progesterone levels measured in unmated adult females sampled weekly throughout the year indicates precisely the seasonal variations in ovarian activity. An anoestrous period appears during the summer and autumn months (Figure 7).

It is postulated that environmental factors, chiefly the length of the day and food availability, may be involved in the control of reproduction. In particular, the annual variation in the availability of autumn mast (mainly acorns and beechnuts) influences the onset of the breeding season; when the amount of mast is large, the season starts early (October) and when it is small, the season starts late (December). Subsequently there can be either 1 or 2 litters/year. The possibility of a second litter seems to be related to the time of weaning. Post-weaning oestrus does not occur if lactation extends to late June or early July, when the anoestrous period begins.

When compared to other ungulates of similar body size, the wild boar's reproductive pattern would seem to be based on good productivity. The seasonal anoestrus, the length of which is markedly influenced by nutritional factors, appears to be the major mechanism controlling natural population growth. However, the short length of gestation, the relatively large litter size and the possibility of a brief anoestrus when food is available make the wild boar the species which has the greatest capacity for reproduction among ungulates.

In modern breeds of domestic swine, descended from the wild form through domestication and intensive selection, the relationship between environment and reproduction has completely disappeared. However, a tendency towards reproductive inefficiency during the summer months still persists, perhaps indicating that the reproductive patterns of the wild boar are not easily eradicated.

5. CONCLUSION

Finally, as we have emphasised in each part of this review, the fundamental characteristic of wild boar and feral hog populations is their great adaptability to environmental conditions. This characteristic permits the use of a large range of rearing systems - extensive or intensive - to which the species can adapt.

6. ACKNOWLEDGEMENT

I am grateful to Dr. J.-P. Signoret for his help in compiling this review.

REFERENCES

Barrett, R., 1978. The feral hog on the Dye Creek Ranch, California. Hilgardia, 46 (9): 283-355. Agric. Sc. Publ. Univ. California.

Beuerle, W., 1975. Freilanduntersuchungen zum Kampf- und Sexualverhalten des europäischen Wildschweines (*Sus scrofa* L.). Z. Tierpsychol., 39: 211-258.

Briedermann, L., 1971. Ermittlungen zur Aktivitätperiodik des Mitteleuropäischen Wildschweines (*Sus s. scrofa* L.). Zool. Garten N.F., Leipzig, 40 (6): 302-327.

Clough, G. and Hassam, A.G., 1970. A quantitative study of the daily activity of the warthog in the Queen Elizabeth National Park, Uganda. E. Afri. Wildl. J., 8: 19-24.

D'Huart, J.-P., 1976. Actogramme journalier de l'Hylochère (*Hylochoerus meinertzhageni* Thomas) au parc national des Virunga, Zaïre. Terre et Vie, 30 (2): 165-178.

Frädrich, H., 1965. Zur Biologie und Ethologie des Warzenschweines (*Phacochoerus aethiopicus* Pallas) unter Berücksichtigung des Verhaltens anderer Suiden. Z. Tierpsychol., 22: 328-393.

Gundlach, H., 1968. Brutsorge, Brutpflege, Verhaltens-ontogenese und Tagesperiodik beim europäischen Wildschwein (*Sus scrofa* L.). Z. Tierpsychol., 25: 955-995.

Hanson, R.P. and Karstad, L., 1959. Feral swine in the south-eastern United States. J. Wildl. Mgmt., 23 (1): 64-74.

Kurz, J.C. and Marchinton, R.L., 1972. Radiotelemetry studies of feral hogs in South Carolina. J. Wildl. Mgmt., 36 (4): 1240-1248.

Martin, J.T., 1975. Movement of feral pigs in North Canterbury, New Zealand. J. Mamm., 56 (4): 914-915.

Mauget, R., 1972. Observations sur la reproduction du Sanglier (*Sus scrofa* L.) à l'état sauvage. Ann. Biol. anim. Bioch. Biophys., 12 (2): 195-202.

Mauget, R., 1978. Seasonal reproductive activity of the European wild boar; comparison with the domestic sow. In: Environmental Endocrinology. Ed. I. Assenmacher and D.S. Farner, Springer Verlag, 79-80.

Mauget, R., 1979. Mise en évidence, par captures-recaptures et radiotracking, du domaine vital chez le Sanglier (*Sus scrofa* L.). Biology of Behaviour, 3: 331-340.

Mauget, R., 1980. Régulations écologiques, comportentales et physiologiques (fonctions de reproduction) de l'adaptation du Sanglier (*Sus scrofa* L.) au milieu. Thèse Sciences, Univ. Tours. 299 p.

Meinhardt, H., 1978. Schwarzwild-Report. Neuman Verlag - Leipzig Radeboel.

Pine, D.S. and Gerdes, G.L., 1973. Wild pigs in Monterey County, California. Calif. Fish and Game, 59 (2): 126-127.

Wood, G.W. and Brenneman, R.E., 1980. Feral hogs movements and habitat use in coastal South Carolina. J. Wildl. Mgmt., 44 (2): 420-427.

DISCUSSION

D.L. Ingram *(UK)*

I wonder if you have had any experience, or have considered the use of an artificial sow? We used one of these at Babraham from immediately after the birth of the piglets until 3 weeks old. It grunted, it was warm, it produced milk and the piglets obviously took to it very well. In fact, when it was withdrawn the piglets showed the same symptoms as if you were withdrawing the mother.

Secondly, would it be practical to make the piglets work some sort of operant response in order to obtain food? Would this be too complicated as a way of enriching their environment?

G. van Putten *(Netherlands)*

I have heard of these experiments but I think it would be very expensive and complicated to replace the sow by an artificial one; in theory we should wean after 3 weeks but the tendency in practice is to wean after 4 or 5 weeks.

I think it would be a good system to let the piglets work for their food - not only piglets in cages but also fattening pigs. A system was designed which forced the pigs to eat as slowly as possible, thus giving them some occupation. In principle, I think it is a good idea in a flatdeck system. However, I still think we should get rid of this system.

J.P. Signoret *(France)*

Are you thinking of some alternative to straw? Are there other possibilities than straw for providing some occupation?

With cattle we have shown that there are very important genetic differences in the requirement for suckling. Has anybody observed genetic differences in the intensity of the urge to suckle amongst the various breeds and strains of pigs?

account, in Figure 5, how much time is spent in such special areas?

R. Mauget

I have not studied this point, but I would say that defaecation usually occurs near the bedding places, as the faeces are found nearby.

H.C. Adler *(Denmark)*

I have some supplementary information on Dr. Mauget's statement that "in numerous herds there persists a trend to a summer reproductive inefficiency". This has also been found in Denmark. At an artificial insemination centre it was found that in August and September a lower non-return percentage occurred than in the rest of the year. The libido of the boars is also lower in these two months.

J.P. Signoret *(France)*

We have also observed this in France.

INFLUENCE OF FIXATION ON THE BEHAVIOUR OF SOWS

K. Vestergaard

Royal Veterinary and Agricultural University,
Department of Forensic and State Veterinary Medicine,
13, Bülowsvej, DK-1870 Copenhagev V, Denmark.

1. INTRODUCTION

In this short paper the scanty evidence available on the behaviour of fixated sows will be evaluated on the basis of the behaviour of unrestrained sows, and potential basic behavioural needs will be discussed. Behavioural needs may be defined as the need to perform:

1) Behaviour which is necessary for the maintenance of a normal physiological state

2) Behaviour which serves other significant functions, e.g. care of skin

3) Behaviour which, when prohibited, can cause frustration and/or damage to the individual.

Any combination of these categories may exist. Sows may be fixated by application of a farrowing crate, by a tie around the neck, a girdle around the chest, or simply by close individual confinement which does not allow the animal to turn round. Fixation may take place during the dry period, around farrowing, or during lactation, and almost any combination of these treatments is found in pig husbandry today. Some behavioural aspects are limited to one or two of the periods while others are general for fixation as such. Accordingly, the present paper deals with an evaluation of behaviour occurring during each of the periods and with a general evaluation of tethering effects.

2. THE DRY PERIOD

Strångby and Gustafsson (1971) examined the time spent lying in different environments during the last 14 days prior to parturition. With increasing restriction they found an increasing amount of lying. The fixated sows spent 92% of the time lying and the free range sows only 72% (day and night). Ekesbo et al. (1978) made similar studies and found that during daytime (6.00 - 18.00 h) fixated sows spent 61% of their time in inaction whereas sows confined in groups of 4 in the 'ALB-system' spent 54% of their day in inactivity. The count for loose housed sows in a more spacious environment was only 40%. Later results (Ekesbo et al., 1979) confirmed the above findings since fixated sows were found to be lying for 68% of the daytime (5.00 - 21.00 h) while the corresponding figure for loose housed sows was only 41%. During the night, both groups of sows spent almost 100% of the time lying. In a small-scale study Vestergaard and Hansen (1977) found that two groups of tethered sows in one observation period spent 51 and 58% of the daytime (5.00 - 1600 h) lying. In the next observation period about 14 days later the corresponding values were 32 and 46% only. In the first period straw was available to both groups but in the second period the group with the lower value was deprived of straw. In the same study sitting like a dog was common, especially during the second observation period, in the group without straw. Sambraus et al. (1978) observed that sitting for 'substantial periods of time' was common in tethered sows kept without straw but rare in loose housed sows. They also noted that the animals were sitting with their heads hanging down or pressed against the stall divisions. Fraser (1975) observed a similar sitting behaviour.

2.1. Influence of straw

When straw is given to fixated sows a substantial part of the time is spent chewing and manipulating the straw (Fraser, 1975) Vestergaard and Hansen, 1977) but in addition there are a number of other effects. Vestergaard and Hansen (1977) found that more time was spent lying when the straw was withdrawn. However, absence of straw increased sitting. Similarly Fraser (1975), who observed sows during an afternoon period, found a decrease in the time spent sitting when straw was given to tethered dry sows which in the preceding period were deprived of straw. Fraser (1975) and Sambraus et al. (1978) found in addition that much time was spent sitting or standing motionless or just sitting and both behaviour patterns were attributed to a lack of physical comfort during lying and accordingly a reluctance to lie down. Repetitive behaviour patterns, including manipulatory activities of mouth and snout are very common in fixated sows. Such behaviour includes nosing and licking bars, floor, trough and chain, biting bars, trough and chain, waving the head from side to side, vacuum chewing and stretching the mouth without any apparent function. In addition there is a great deal of drinking (play drinking?) and pressing or rubbing the snout in the trough or against the bars to be seen during standing and sitting (Fraser, 1975; Sambraus et al., 1978). In general, all these patterns can be reduced by the provision of straw (Fraser, 1975; Vestergaard and Hansen, 1977) (Table 1).

TABLE 1

THE INFLUENCE OF STRAW ON SELECTED BEHAVIOUR PATTERNS OF FIXATED DRY SOWS. THE SCORES REPRESENT THE MEAN NUMBER OF MIN EACH PATTERN OCCURRED DURING A 40 MIN OBSERVATION (VESTERGAARD AND HANSEN, 1977).

	Manipulation of straw	Straw chewing	Bar biting	Head waving	Waving bite	Sit or stand motionless
- straw	0	0	19	15	12	19
+ straw	35	35	1	0	1	3
Significance of difference	x	x	x	xx	NS	NS

2.2. Straw availability

The ways in which straw exerts an effect have been examined by Fraser (1975) who speculated that straw might function by compensating for the lack of bulk in the diet, reducing 'boredom', and providing a more comfortable bedding. These 'dietary', 'recreational' and 'bedding' aspects were examined by providing the straw in three different ways:

1) chopped straw mixed with the food, i.e. straw available only as dietary bulk

2) loose straw in the food trough for recreational purposes and as a bulk but not available for bedding

3) a large amount of loose straw both in the trough and on the lying area ('full bedding').

These different conditions were compared with each other and with the condition in which no straw was available (Table 2). The chopped straw in the food resulted in more lying and less standing than in the case without straw but it failed to influence the other behaviour patterns mentioned above. Sitting was reduced by 'full bedding' when compared to the situation where chopped straw was provided in the food. Nosing and licking of bars, floor, trough and chain were reduced somewhat by the provision of straw in the trough and reduced much more by 'full bedding'. Chopped straw in the food had no effect on these patterns. 'Vacuum' chewing was reduced by 'full bedding' when compared to the other treatments, between which there was no difference. Mouth stretching was reduced by 'full bedding' in comparison to the situation where there was no straw. In comparison to the situation without straw, head waving was reduced by 'full bedding' and by loose straw in the trough but chopped straw in the food had no effect. Less head waving occurred with 'full bedding' than with chopped straw in the food. Thus, the results clearly indicate that straw exerts its effect in the three ways suggested. The 'dietary' effect was mainly that of increased lying and reduced standing, whilst the oral and other stereotyped behaviours were reduced by loose straw, especially when 'full bedding' was provided. The straw and the way in which it was presented did not influence aggression but provision of loose straw and 'full bedding' reduced the amount of nosing and licking of neighbour pigs. Comfort movements occurred, but they were unaffected by the different treatments.

2.3. Evaluation

In general, an evaluation of the severity of the behavioural changes due to fixation seems difficult on the basis of present knowledge. Changes in the frequency of standing, lying and sitting are especially difficult to evaluate because nothing is known of the normal range of frequency of these patterns. Standing, and sometimes sitting, have often been considered a sign of activity

TABLE 2

MEAN SCORES (NUMBER OF MIN OUT OF 80 IN WHICH THE BEHAVIOUR PATTERN WAS OBSERVED) FOR 4 POSTURES AND 11 ACTIVITIES WHEN STRAW WAS GIVEN AS FULL BEDDING (F), IN THE TROUGH (T), CHOPPED AND MIXED WITH FOOD (C) AND WHEN NO STRAW WAS GIVEN (N) (FRASER, 1975).

Behaviour	Experimental treatment				Statistical comparison
	Full (F)	Trough (T)	Chopped (C)	No straw (N)	
Postures:					
Lying	74.8	74.2	84.6	46.2	$C > N^{xx}$. $F, T > N^{x}$.
Standing	125.3	125.0	105.5	144.2	$C < N^{xx}$. $C < T^{x}$.
Sitting	2.2	4.6	14.1	12.3	$F < C^{x}$.
Standing or sitting motionless	1.8	18.0	38.2	35.3	$F < T, C, N^{xx}$. $T < C, N^{x}$.
Main activities (as percentage of time spent standing)					
Chewing and manipulating straw	92.1	61.5	13.1	1.8	$F > T > C > N^{xx}$.
Nosing and licking bars, floor, trough and chain	3.1	10.7	33.6	30.7	$F < T < C, N^{xx}$.
Biting bars, trough and chain	5.1	11.3	37.2	42.5	$F < T < C, N^{xx}$.

TABLE 2 (CONTD)

Behaviour	Experimental treatment				Statistical comparison
	Full (F)	Trough (T)	Chopped (C)	No straw (N)	
Less common activities:					
Rubbing on walls	2.4	2.5	2.9	2.3	-
Scratching with hind leg	1.3	1.0	1.2	1.2	-
Head shaking	2.4	4.5	4.4	4.4	-
Nosing and licking neighbour	0.1	0.5	1.0	1.7	$F < N^{xx}$.
Biting neighbour	1.3	1.2	1.0	1.2	-
Head waving	0.9	6.1	11.1	20.8	$F, T < N^{xx}$. $T < C^{x}$.
Chewing (in vacuum)	1.4	3.8	11.9	18.9	$F < T, C, N^{xx}$. $T < N^{xx}$.
Stretching mouth	0	1.7	3.5	3.5	$F < N^{x}$.

but such patterns cannot be directly compared to the walking or running of the unrestrained sow since only these patterns imply exercise. It should be noted, however, that the studies available, although few, are consistent in that fixation increases lying and sitting and that straw reduces sitting and increases lying in fixated dry sows.

2.4. Stereotyped behaviour

Stereotyped behaviour generally develops in animals which have been exposed at an early age to social deprivation and/or a deficient environment, and in many cases this behaviour develops from escape movements (Hinde, 1972). Such behaviour is common in wild animals in zoological gardens (Morris, 1964). It is indicative of fear in the domestic fowl as it can be reduced during frustration by the administration of a fear-reducing drug (pacitran) (Duncan and Wood-Gush, 1974). Thus, the performance of stereotyped behaviour may be a sign of fear, but it has also been suggested that it serves to reduce fear or simply to increase sensory input (Morris, 1964). Stereotyped behaviour may be divorced from the conditions which originally caused it (Hinde, 1972; Ödberg, 1978) but that does not seem to be totally true in the case of the fixated sow since such behaviour increases within a few days of the removal of straw and decreases as soon as straw is replaced. However, there have been no studies on the significance of such behaviour in connection with the well-being of mammals. From operant studies on pigs there has been no recorded stereotyped behaviour, even though adverse procedures were involved (Baldwin and Stephens, 1973; Dantzer and Baldwin, 1974; Mormède and Dantzer, 1977), but it is possible that the short duration of these experiments did not allow sufficient time for stereotyped behaviour to develop. On the other hand, frustration may lead to stereotyped behaviour in other mammals and in birds (Ödberg, 1978). There is now, therefore, a great need for basic studies on the significance of the occurrence and development of these behavioural patterns in relation to the well-being of sows.

3. FARROWING SOWS

Around farrowing, fixation of sows (without a tie) in one study resulted in lying during 90 - 100% of the time. The loose housed sows to which the fixated sows were compared were found to be lying 78 - 80% of the time during the same period (Strångby and Gustafsson, 1971). However, the effect of conditions prior to parturition could not be evaluated in that study since only the fixated sows had been fixated during the preceding dry period, whilst the loose sows had been kept loose in groups. Again, as with dry sows, the significance of the changes of duration of the postures is unknown. Similarly, Baxter and Petherick (1980) recorded much more activity in the unrestrained but isolated sow during the 24 - 36 h prior to parturition. Activity increased to 10 - 15 times the normal level and nest building patterns became frequent. These patterns were:

1) gathering by mouth
2) depositing in one area of the pen
3) burrowing in the pile of straw whilst on front knees and displaying strong rooting movements
4) pawing in the straw.

This behaviour produced a large, cup-shaped nest about 175 cm in diameter. In the conventional farrowing crates without any litter the level of activity did not change much, but intense, stereotyped bar-gnawing and pawing behaviour were seen together with strenuous attempts to escape from the crate. Furthermore, the animals showed 'copious frothing at the mouth' and pupillary dilation. Nest-building such as that observed in the loose sows is well-known from observations made of free range sows and the wild boar (Signoret et al., 1975). The behaviour of the restrained sows, however, was interpreted as displacement nest-building behaviour and all the above mentioned symptoms were considered to be evidence of stress due to frustration of nesting activities. Scratching (pawing?) (and rooting the floor) has been explained as a frustration response to a 'time out' procedure in an operant condition study of pigs (Dantzer and Baldwin, 1974) but in that situation there was no bradycardia as was found when the pigs were subjected to punishment of the operant response by electric shocks. Farrowing of fixated sows tends to be of a longer duration and involve more complications than that of loose sows (Strångby and Gustafsson,1971; Bäckström, 1973).

4. LACTATING SOWS

The postures of lactating fixated sows have been studied by Strångby and Gustafsson (1971). In general, there is little difference in the time spent by fixated and loose sows in lying during the night period, but during daytime fixated sows initially show a large amount of lying, as in the farrowing period, followed by a drop to about 55% at weaning in the seventh week. Loose housed sows, especially those which have been outside in the preceding dry period, tend to maintain a stable amount of lying at about 70% of the day (Figure 1).

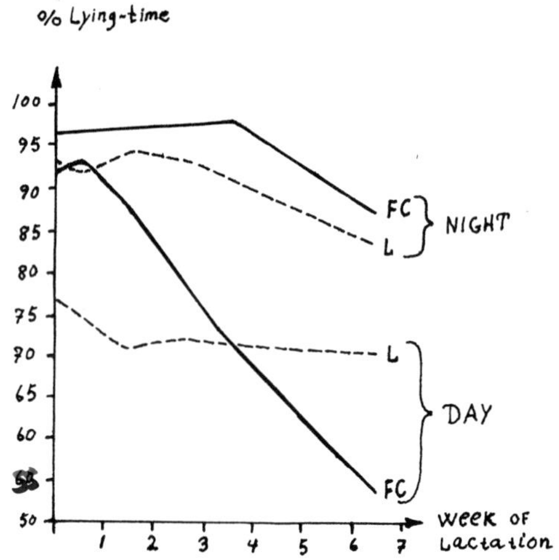

Fig. 1. Percentage lying time of fixated sows (FC) and loose sows (L) during daytime (6.00 - 18.00 h) and night time (18.00 - 6.00 h) (Strångby and Gustafsson, 1971).

In fixated sows, lying on the belly was observed more frequently than in loose sows and from week three this posture increased markedly in the fixated sows but decreased in the loose sows (Figure 2). This finding was interpreted as an attempt on the part of the fixated sow to avoid excessive lactations, but these attempts generally failed since lactation was observed much more frequently in the fixated sows than in the loose sows during the last three weeks of lactation (Figure 3).

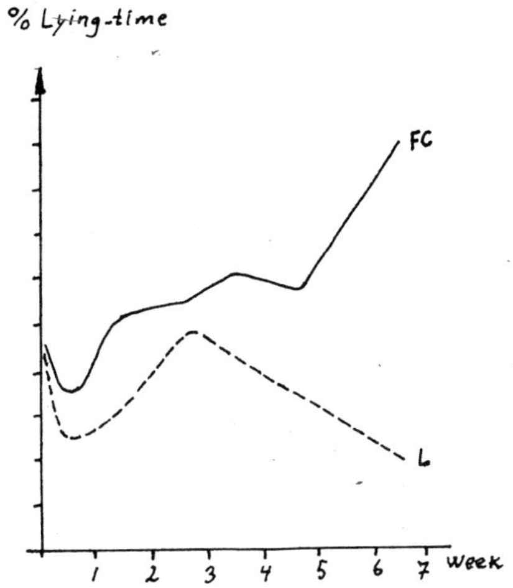

Fig. 2. Percentage of time (day and night) spent lying on the belly in fixated sows in farrowing crates (FC) and in loose sows (L) (Strängby and Gustafsson, 1971).

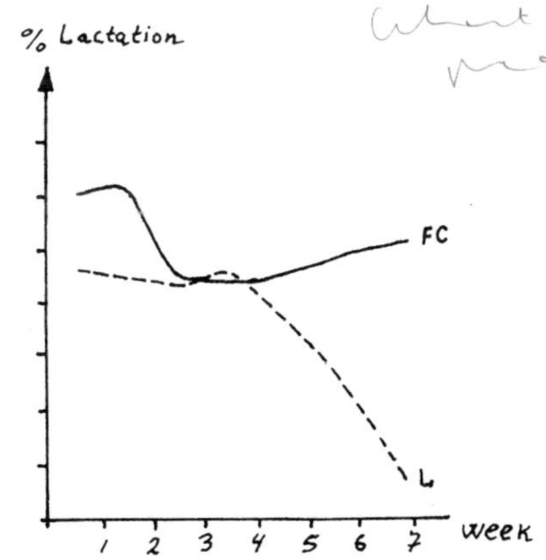

Fig. 3. Frequency of lactation in sows in farrowing crates (FC) and in loose sows (L) in individual pens (Strängby and Gustafsson, 1971).

In the above studies the effects of bedding and type of floor could not be evaluated. However, Gravås (1979) compared different floors in farrowing crates and found that sows kept on rubber mats spent 86% of the time (day and night) lying during the period between day 3 and day 17 of lactation. The corresponding values for epoxy and concrete floors were 81% and 75% respectively. Gravås postulated that a large amount of lying was a basic requirement of sows with newborn piglets. The values for rubber mats were very similar to those observed by Strångby and Gustafsson (1971) in loose housed sows provided with straw bedding. Sitting is much more common in fixated lactating sows than in loose lactating sows (Strångby and Gustafsson, 1971). Again, the link between such posture frequencies and well-being is uncertain.

4.1. Potential behaviour problems in fixated sows

In addition to the behavioural changes discussed above, changes in comfort and eliminative behaviour may occur during all periods of fixation. In particular, there are limited possiilities for performing normal temperature regulatory behaviour. Normally pigs huddle when it is cold, and when it becomes hot they seek shade or wet their skin by lying in wet places, bathing or wallowing (Heitman et al., 1962; van Putten, 1978). Thus, pigs are typical behavioural temperature regulators and it is well known that they lack thermo-regulatory sweating and effective panting (Mount, 1979). As a consequence, fixated sows may fail to regulate their temperature according to their need. During hot periods play drinking may be caused by an attempt to spout water on the lying area (van Putten, 1978) but such adjustments probably have only a limited value, especially where water is only available around feeding time. Since basic physiological mechanisms are involved, it seems that behavioural temperature regulation may be of essential significance to the well-being of the animal. However, although there are a number of papers dealing with the detrimental effects of high temperature on reproduction (Love, 1978), nothing is known of its effect on the well-being of the animals.

4.2. Other comfort movements

There are a number of other comfort movements in the pig (van Putten, 1978) which may be influenced by fixation, but no evidence of such effects has been presented, with the exception of rubbing on the walls, scratching with the hind leg and head shaking which have been recorded in fixated dry sows (Fraser, 1975).

4.3. Frustration

Defaecation is often preceded by bar biting (Fraser, 1975; Vestergaard, unpublished) which may be a sign of frustration. According to Sambraus et al. (1978) most fixated sows move forward prior to defaecation, which results in dirty lying places. This behaviour may be interpreted as an attempt to perform the normal excursion away from the resting area before defaecation takes place.

During fixation movements in contact with the equipment may result in traumatic injuries (Ekesbo, 1973), and leg weakness may be aggravated due to the restraint on movement. From a study of boars it has been found that exercise reduces the incidence of leg weakness of the forelimbs (Perrin and Bowland, 1977). In Denmark, gilts are often kept loose since it is generally found that they tend to develop leg weakness if fixated. Fixated lean sows kept without strawbedding frequently develop decubitus ulcers in the shoulder region (Sambraus et al., 1978) and this may partly be dependent on the increased lying time resulting from fixation.

5. CONCLUSION

In conclusion, fixation results in a number of behavioural changes besides a reduction in locomotor activity. Some of the changes, especially those concerning postures, are difficult to evaluate from the point of view of welfare, but others, such as the increased amount of stereotyped behaviour, are indications of stress. Some of the behaviour patterns which are curtailed by fixation may represent essential behavioural needs since basic physiological mechanisms are affected or because there is frustration. The absence of straw tends to aggravate the situation. Fixation often inflicts traumatic injuries, but the incidence is probably greatly related to housing details. In one study fixation resulted in increased adrenal weight (Jensen et al., 1970) but there is no additional information available concerning physiological stress responses due to fixation. In future studies such measures should be taken in combination with observation of behaviour.

REFERENCES

Baldwin, B.A. and Stephens, D.B., 1973. The effects of conditioned behaviour and environmental factors on plasma corticosteroid levels in pigs. Physiology and Behaviour, 10: 267-274.

Baxter, M.R. and Petherick, J.C., 1980. The effect of restraint on parturition in the sow. International Pig Veterinary Society. Congress proceedings, Copenhagen, p.4.

Bäckström, L., 1973. Environment and animal health in piglet production - a field study of incidences and correlations. Acta Veterinaria Scandinavica Suppl. 41: 1-240.

Dantzer, R. and Baldwin, B.A., 1974. Changes in heart rate during suppression of operant responding in pigs. Physiology and Behaviour, 12: 385-391.

Duncan, I.J.H. and Wood-Gush, D.G.M., 1974. The effect of a rauwolfia tranquillizer on stereotyped movements in frustrated domestic fowl. Applied Animal Ethology, 1: 67-76.

Ekesbo, I., 1973. Animal health, behaviour and disease prevention in modern Swedish animal husbandry. The Veterinary Record, 93: 36-39.

Ekesbo, I., Jensen, P. and Högsved, O., 1978. Nygammal typ av sinsuggehålning - några etologiska data. Svensk Veterinärtidning, 30: 845-848.

Ekesbo, I., Jensen, P. and Lock, R., 1979. Några exempel på beteendeförändringer vid fixering av sinsuggor. Svensk Veterinärtidning, 31: 315-319.

Fraser, D., 1975. The effect of straw on the behaviour of sows in tether stalls. Animal Production, 21: 59-68.

Gravås, L., 1975. Behavioural and physical effects of flooring on piglets and sows. Applied Animal Ethology, 5: 333-345.

Heitman, H. Jr., Hahn, L., Bond, T.E. and Kelly, C.F., 1962. The effects of modified summer environment on swine behaviour. Animal Behaviour, 10: 15-19.

Hinde, R.A., 1972. Animal Behaviour. McGraw-Hill.

Jensen, A.H., Yen, J.T., Gehring, M.M., Baker, D.H., Becker, D.E. and Harmon, B.G., 1970. Effects of space restriction and management on pre- and post-puberal response of female swine. Journal of Animal Science, 31: 745-750.

Love, R.J., 1978. Definition of a seasonal infertility problem in pigs. The Veterinary Record, 103: 443-446.

Mormède, P. and Dantzer, R., 1977. Experimental studies on avoidance behaviour in pigs. Applied Animal Ethology, 3: 173-185.

Morris, D., 1964. The response of animals to a restricted environment. Symposium of the Zoological Society of London, 13: 99-118.

Mount, L.E., 1979. Adaptation to thermal environment. Edward Arnold, London.

Odberg, F.O., 1978. Abnormal behaviours: (Stereotypies). In: 1st World Congress on Ethology Applied to Zootechnics, Madrid, 475-480.

Perrin, W.R. and Bowland, J.P., 1977. Effects of enforced exercise on the incidence of leg weakness in growing boars. Canadian Journal of Animal Science, 57: 245-253.

van Putten, G., 1978. Comfort behaviour in pigs: informative for their well-being. In: The ethology and ethics of farm animal production, Tierhaltung, 6: 70-77.

Sambraus, H.H., Sommer, B. and Kräusslich, H., 1978. Verhaltung von Sauen in verschiedenen Haltungssystemen. In: 1st World Congress on Ethology applied to Zootechnics, Madrid, 99-102.

Signoret, J.P., Baldwin, B.A., Fraser, D. and Hafez, E.S.E., 1975. The behaviour of swine. In: The Behaviour of Domestic Animals, Baillière Tindall, London, 295-329.

Strangby, G. and Gustafsson, B., 1971. Beteendestudier i suggstallar. Aktuelt från Landbrukshögskolan, no 171. Teknik, 14: 1-42.

Vestergaard, K. and Hansen, L.L., 1977. Adfaerd hos draegtige søer i enkeltbokse med og uden halm. (Stencil), 1-10.

DISCUSSION

G. van Putten *(Netherlands)*

Did you discriminate in the method of fixating the sows - you only distinguished between fixated and non-fixated sows.

K. Vestergaard *(Denmark)*

In the studies that I made myself we only had fixation due to confinement, except during the dry period when we had tethering.

W.T. Truyen *(Netherlands)*

Is there any information on the disadvantages of grouping sows, instead of tethering them, in regard to the pattern of fighting? There is also more chance of wounds on legs which can cause infections and helminth infestation, among other things.

K. Vestergaard

Unfortunately, there is very little information available, although I have begun investigating this aspect. Mainly the effect of straw on the fixated sows has been investigated. There is very little information about social behaviour, contrasting loose and fixated sows.

R.A. Dantzer *(France)*

I would like to mention that drinking, by itself, is certainly not a displacement activity but is related, as you have said, to thermoregulatory behaviour. However, playing with drinking water is another matter, and this might be a displacement activity. In some recent experiments, we have been trying to induce psychogenic polydipsia, which is common behaviour in rats submitted to conflict situations. However, this does not work with pigs.

I. Ekesbo *(Sweden)*

I would like to give one figure concerning social interaction in confined and loose sows. At my institute we have studied about 20 sows, in two groups, and we found 1 216 social interactions within the 20 loose sows, and only 236 social interactions within the 16 confined sows. This gives a quantitative idea for the social interactions in the two different keeping regimes for dry sows. With regard to quality, we have found that some social behaviour patterns found in loose sows do not occur at all amongst confined sows, and some occur only in part.

PHYSIOLOGY AND BEHAVIOUR OF YOUNG PIGS IN RELATION TO THE ENVIRONMENT

D.L. Ingram

ARC Institute of Animal Physiology, Babraham, Cambridge, UK.

ABSTRACT

Thermal preferences (radiant temperature, air movement, nature of the floor) and light preferences have been studied in the laboratory and under natural conditions. The animals' response to lighting and circadian variations in the demand for heating have also been examined.

These studies help us to understand the changes in behaviour induced by the environment under husbandry conditions.

1. INTRODUCTION

The influence of the environment on the behaviour and physiology of the young pig has been studied at the Institute of Animal Physiology under a number of headings. Initially interest focused on thermal effects with respect to the conditions which favoured growth, and subsequently interest has broadened to include food and water intake, lighting and social behaviour.

2. THERMAL INFLUENCES

2.1. Temperature

In very young pigs Mount (1963) demonstrated that the temperature selected in a thermocline was close to the critical temperature, although the spread of choices was wide. Later studies by Baldwin and Ingram (1967, 1968a, 1968b) employed the technique of operant conditioning in which weaner pigs learned to operate a switch for a reinforcement consisting of a short burst of radiant heat. Under these conditions the number of reinforcements obtained in unit time decreased as the ambient temperature increased and fell to near zero at the critical temperature. Within a restricted range it could also be shown that the rate at which reinforcements were obtained varied with the intensity of heating and its duration. Moreover pigs on a low energy intake obtained more reinforcements than those on a high intake. The animals' natural thermal preference is thus close to the conditions which are favourable to growth and efficient use of food (Mount, 1979; Close et al., 1978). If the task of obtaining heat was made more difficult by making it necessary to operate the switch several times for one reinforcement, or if the intensity of heating was very low the relation between the ambient temperature and the rate of obtaining reinforcements was lost (Ingram, 1975).

Under natural conditions pigs showed no tendency to select the region of highest radiant heat, and took shelter only when the ambient temperature was well below the critical temperature (Ingram and Legge, 1970a). It must therefore be concluded that although the pig has a strong tendency to select conditions which are near optimal for growth it will do so only under very restricted circumstances.

2.2. Air Movement

Another aspect of the environment which affects heat loss is air movement which is particularly important for the pig because it has very little external insulation (Mount and Ingram, 1964; Ingram et al., 1975). It was found that pigs will also make a response consisting of pressing a switch when the reinforcement was a short respite from a draught caused by a fan (Baldwin and Ingram, 1967; 1968a). Again the number of reinforcements obtained correlated well with the ambient temperature and there was a sharp fall in the number of responses above the critical temperature. The pigs did not, however, adjust the rate of reinforcement in strict relation to the duration of the respite, nor to the force of the draught. It was concluded that the aversive nature of high air movement may be related to other factors besides its effect on heat loss. This conclusion was strengthened by studies in which metabolic rate was measured while the pig made responses to turn the fan off, and it was found that there was no measurable saving in oxygen consumption as a result of the pigs' behaviour (Ingram, 1975). Nevertheless under natural conditions outdoors pigs always selected the region which had the lowest air movement (Ingram and Legge, 1970a). It may therefore be possible to influence a pig's choice of position by differences in air movement which need not be excessive.

2.3. Nature of the floor

The pig's comfort in relation to heat loss can also be influenced by the nature of the floor and the presence of bedding. A concrete floor, even if it is insulated from the ground has a high thermal capacity and if a pig lies down on it heat is rapidly conducted away. Under such circumstances the pig adapts a tense position lying on its limbs. When the animal is placed on a slab of expanded polystyrene at the same ambient temperature the animal relaxes, and even though the area of contact with the floor is greater, heat loss is decreased (Mount, 1967; 1968). This point was systematically investigated by Stephens (1971). For the newborn pig at 10°C the effect of providing straw was equivalent to increasing the air temperature to 18°C. This very considerable advantage of straw is due to its reduction in both conduction and convective heat loss and represents a most effective means of increasing both comfort and efficiency of food utilisation in the pig.

2.4. Thermal, food and water intake relationship

The thermal environment also influences the quantity of food and water demanded by animals: if energy intake is restricted growth must be retarded progressively as the temperature falls. The length of time for which the environmental conditions must be altered in order to influence voluntary intake of food was studied by Ingram and Legge (1974). Weaner pigs were kept at 25°C and fed once a day in a cage where they operated a switch for a reinforcement of a few grams of food. The temperature of the room in which the feeding cage was housed was either 10, 25 or 35°C and the animals were left until the switch was no longer operated. Under these conditions food intake was reduced at 35°C by comparison with 25°C, but at 10°C there was no change.

In a second series of studies the pigs were exposed in the cage to 10, 25 or 35°C for 2 or 4 h before the switch was made available. Food intake was then increased at 10°C, but had almost returned to normal at 35°C. Similarly it was found that after several days exposure to 10 or 35°C after a period of 25°C food intake was increased at 10°C, but at 35°C it was hardly affected.

The physiological nature of the signals involved in these changes are not readily apparent. Using the operant conditioning technique, feeding took some 2 h to complete and it is known that any change in metabolic rate in relation to temperature is complete after only 30 min. It therefore seems unlikely that it was a change in metabolic rate which stimulated the animal to obtain more food. A possible clue stems from experiments in which the hypothalamic temperature was changed by means of a thermode during feeding. It was found that heating caused an increase in food intake and cooling a decrease (Ingram, 1968). Since it is also known that body temperature increases temporarily on transfer from a hot to a cold environment, and decreases on transfer from cold to hot (Ingram and Legge, 1970b), it is possible that these changes in body temperature are instrumental in changing food intake. In any event the indication is that a change in the ambient temperature for only a short while before feeding can influence the demand for food.

Water intake in pigs can also be influenced by the environmental temperature (Mount et al., 1971), but the changes induced thermally are small compared with those due to osmotic stimuli (Ingram and Stephens, 1979).

3. LIGHTING AND CIRCADIAN RHYTHMS

Most animals display circadian rhythms in physiological variables, and in the presence of light/dark cycles these can become entrained in a 24 h cycle. Thus diurnal animals are more active by day and their body temperature begins to decrease at the end of the day and increase again towards morning. The pig by contrast has no marked circadian rhythm of body temperature. When housed singly in a cage and fed once a day, the core temperature increased after the meal and then fell again over the next 12 hours or so, irrespective of the time of day at which feeding occurred. If the animals were fed *ad libitum* there was again no circadian rhythm. A rhythm in body temperature was seen only in caged animals which had been deprived of food over 4 days, or in groups of pigs kept in a hut with access to a small paddock (Ingram and Legge, 1970b; Ingram and Mount, 1973). The demand for heat using the technique of operant conditioning did display a circadian rhythm but this did not appear to be related either to the presence of a light/dark cycle, or to the time of feeding (Ingram et al., 1975).

3.1. Light/dark cycle

These findings prompted the investigation of rhythms of activity and food and water intake in relation to light/dark cycles. Groups of four pigs were housed together in a pen which was completely isolated from all outside sounds and lighting. They were provided with food and water *ad libitum* and the amounts taken were recorded automatically every hour. Cleaning was carried out at irregular times of the day at irregular intervals of 2, 3 or 4 days. For two weeks the light was on for 12 h in each 24, and then for the next two weeks the animals were in continuous light. Under these conditions most activity occurred during the period of light, and most food was taken towards the end of the light period. Both activity and feeding, however, continued into the dark period and there were very few hours in which no food was taken. There was no tendency for discrete meals to be taken with intervals between. In conditions of constant light, the rhythms of activity were soon diminished and in some instances completely lost. Pigs exposed individually under the same conditions displayed less well defined rhythms, and one pig was consistently active in the dark (Ingram et al., 1980). These studies have led to the view that the pig's behaviour is such that the animal is likely to adapt to a wide variety of feeding and lighting regimens. Indeed this may partly explain why it can be kept under intensive husbandry conditions.

3.2. Illumination

The question of the animal's preference for illumination has been further tested by Baldwin and Meese (1975). They arranged for pigs to be able to break an infra red beam with their snout and so hold a light on in the room for as long as the beam was broken. As soon as the snout was removed from the beam the light went out. A second identical beam switch which did not operate the light served as a control. It was found that the pigs displayed a very strong preference for using the switch which operated the light by comparison with the control switch, but the total amount of time for which the light was kept on was less than 1% of the duration of the experiment. In a second study the beam switches were arranged so that breaking the beam turned the light on and it stayed on until a second beam was broken to turn it off. In this study the pigs had the light on for 72% of the time, although they did have a period of darkness during most hours. There is thus reason to believe that the pig has a strong preference for light, but the effort which it is prepared to make in order to be in a light room is rather restricted. Such techniques should prove very useful in determining precise preferred levels of illumination corresponding to the way in which the preferred temperature can be determined.

4. LONG TERM EFFECTS OF ENVIRONMENT

The longer term effects of environment have been less completely studied at Babraham and only recently have the effects of early weaning on behaviour and brain chemistry been considered. Prolonged exposure of pigs to hot or cold conditions has been shown by operant conditioning to affect their demand for heat (Baldwin and Ingram, 1968b). The effects were not, however, as might have been expected since pigs which had been kept in the cold demanded more heat than those acclimatised to the heat. A further study of the problem (Heath and Ingram, 1978; Heath, 1979) raised pigs from shortly after birth at 10 or 35°C and found that the tendency for the pigs from the cold to demand more heat persisted even if the animals were kept at a neutral temperature for 24 h before being tested.

Such observations suggest that early experience may have profound effects on control mechanisms. Evidence that the environmental condition in the neonatal period can in fact change the brain chemistry has recently been obtained at Babraham in studies of early weaning (Mann and Sharman, personal communication). In these experiments animals which had been removed from the sow at 12 - 14 days and kept in separate cages were compared with littermates left with the mother. At the end of 10 days the concentrations of catecholamine metabolites and the synthesising enzyme tyrosine hydroxylase in the caudate nucleus fell below those found in littermates reared by the sow. Studies of this kind will it is believed lead us to better understanding of the basic physiological mechanisms which underline the changes in behaviour which may be induced under particular husbandry conditions.

REFERENCES

Baldwin, B.A. and Ingram, D.L., 1967. Behavioural Thermoregulation in pigs. Physiol. Behav. 2, 15-21.

Baldwin, B.A. and Ingram, D.L., 1968a. Factors influencing behavioural thermoregulation in pigs. Physiol. Behav. 3, 409-415.

Baldwin, B.A. and Ingram, D.L., 1968b. The effect of temperature acclimation and food intake on behavioural thermoregulation in pigs and mice. Physiol. Behav. 3, 395-400.

Baldwin, B.A. and Meese, G.B., 1977. Sensory reinforcement and illumination preference in the domestic pig. Anim. Behav. 25, 497-507.

Close, W.H., Mount, L.E. and Brown, D., 1978. The effects of plane of nutrition and environmental temperature on the energy metabolism of the growing pig. 2. Growth rate, including protein and fat deposition. Brit. J. Nutr. 40, 423-431.

Heath, M.E., 1980. Effect of rearing temperature on the thermoregulatory behaviour of pigs. Behav. Neural Biol. 28, 193-202.

Heath, M.E. and Ingram, D.L., 1978. Physiological effects of rearing young pigs in hot and cold environments. J. Physiol., Lond. 284, 8-9P.

Ingram, D.L., 1968. The effect of heating and cooling the hypothalamus on food intake in the pig. Brain Res. 11, 714-716.

Ingram, D.L., 1975. The efficiency of operant thermoregulatory behaviour in pigs as determined from the rate of oxygen consumption. Pflügers. Arch. ges. Physiol. 353, 139-150.

Ingram, D.L., Heal, J.W. and Legge, K.F., 1975. Heat loss from unrestrained pigs in an outdoor environment. Comp. Biochem. Physiol. 50, 71-76.

Ingram, D.L. and Legge, K.F., 1970a. The thermoregulatory behaviour of young pigs in a natural environment. Physiol. Behav. 5, 981-987.

Ingram, D.L. and Legge, K.F., 1970b. Variations in deep body temperature in the young unrestrained pig over the 24 h period. J. Physiol., Lond. 210, 989-998.

Ingram, D.L. and Legge, K.F., 1974. Effects of environmental temperature on food intake in growing pigs. Comp. Biochem. Physiol. 48, 573-581.

Ingram, D.L. and Mount, L.E., 1973. The effects of food intake and fasting on 24 h variations in body temperature in the young pig. Pflügers. Arch. ges. Physiol. 339, 299-304.

Ingram, D.L., Walters, D.E. and Legge, K.F., 1975. Variations in behavioural thermoregulation in the young pig over 24 hour periods. Physiol. Behav. 14, 689-695.

Ingram, D.L., Walters, D.E. and Legge, K.F., 1980. Variations in motor activity and food and water intake over 24 hour periods in the pig. J. agric. sci. (in press).

Ingram, D.L. and Stephens, D.B., 1979. The relative importance of thermal, osmotic and hypovolaemic factors in the control of drinking in the pig. J. Physiol., Lond. 293, 501-512.

Mount, L.E., 1963. The environmental temperature preferred by the young pig. Nature, Lond. 199, 122-123.

Mount, L.E., 1967. The heat loss from new-born pigs to the floor. Res. vet. sci. 8, 175-186.

Mount, L.E., 1968. The Climatic Physiology of the Pig. Edward Arnold, London.

Mount, L.E., 1979. Adaptation to Thermal Environment. Edward Arnold, London.

Mount, L.E., Holmes, C.W., Close, W.H., Morrison, S.R. and Start, I.B., 1971. A note on the consumption of water by the growing pig at several environmental temperatures and levels of feeding. Anim. Prod. 13, 561-563.

Mount, L.E. and Ingram, D.L., 1964. The effect of ambient temperature and air movement on localized heat loss from the pig. Res. vet. sci. 6, 84-91.

Stephens, D.B., 1971. The metabolic rates of newborn pigs in relation to floor insulation and ambient temperature. Anim. Prod. 13, 303-313.

DISCUSSION

I. Ekesbo *(Sweden)*

You showed us that in your last experiment the animals were kept at +10 and +30°C, and revealed the differences not only in the figures, but also in the behaviour of the animals. But what was the lying area like? Was it a very barren environment or was there straw?

D.L. Ingram *(UK)*

The animals were kept in cages on wire mesh. If one wants to look at the effects of environment, then providing straw or allowing two animals to get together destroys the object of the exercise. So, it was a very barren environment.

G. van Putten *(Netherlands)*

You stated that animals kept in a rich environment have lower temperature demands and wind velocity is not so important. However, in the experiments of Dr. Baldwin, the animals were kept in a very poor environment and yet you refer to these experiments as valid. Do you not think that the environment of the individually kept pigs was so poor that their demands were much higher than they would have been if kept in a socially richer environment?

D.L. Ingram

If you are talking about the extent to which the animals will make an operant response, you are right. Thermally, pigs will demand heat in a restricted environment. However, it is not that they do not need thermoregulation in a rich environment; they simply don't use it. The heat loss from pigs kept outside in environments of about 10°C with the rain and the wind, is really quite considerable. Obviously, an animal in a cage is in a very artificial environment but I think the order of the response remains the same - the animal will still turn off the draught at a low temperature and will ignore the switch when the temperature is high. I think you get the answer to whether the pig finds this uncomfortable, and whether he finds it more uncomfortable at other temperatures. The enrichment of the environment is a very complicating factor - an environment could be very rich and yet physiologically very unsuitable for the pig.

T. Hanrahan *(Ireland)*

The last point in your paper, Dr. Ingram, was about environment affecting brain chemistry. Are there examples of this happening in other animals, and what effect does this have on the subsequent behaviour of the animals? If their brain chemistry is affected, then in what way is behaviour altered?

D.L. Ingram

The experiments were first done on guinea pigs in relation to chewing behaviour. If guinea pigs are isolated in a poor environment they tend to lick the bars; pigs do the same sort of thing, as well as tail-biting. Although these are early experiments, we think that the change in the concentration of the catecholamines in the caudatus nucleus may be correlated with this vice behaviour.

R.G. Buré *(Netherlands)*

The dark-light situation is one of the possibilities that could have an influence on the rhythm of activity. Are there any other things that are important to this rhythm, for instance feeding-time? Could feeding-time be as, or more, important, or is the dark-light situation the most important factor in that rhythm of activity?

D.L. Ingram

I have not tried feeding the animals at different times. These animals were fed with a device which provided them with an open bowl from which we could measure by means of a weighing device how much food had been taken, and replace it remotely.

I would guess that the time of feeding would also change the activity rhythm. In this case, the light was the only marked variable. If you try cycling the daylight at the same time and make the temperature very hot, about 35 - 40°C, then the activity is decreased and you can make the animals behave as nocturnal animals.

ANIMAL WELFARE IN AGRICULTURE - A CHALLENGE FOR THE ETHOLOGY

H. Bogner

Bayer. Landesanstalt für Tierzucht, Prof. Dürrwaechter-Pl. 1,
D - 8011 - Grub/p. Poing, Federal Republic of Germany.

1. INTRODUCTION

The European Agreement for animal protection in agricultural husbandry of March 10th 1976 has already become law in several member states of the European council.

According to this law each animal, especially in modern intensive housing systems, has to be stalled, nourished and treated in a manner appropriate to the species. Therefore, its state of development adaptability and domestication, as well as its physiological and ethological requirements have to be taken into consideration in accordance with present experience and scientific knowledge.

All regulations of this agreement are aimed at preventing animals from harm and suffering. Contrary to the German Law for animal protection of 1972,'animal well-being' is nowhere directly mentioned in the European agreement. It is also accepted (Art. 4,2) that animals can be kept regularly or permanently tied, chained or penned so far as their minimum space requirements are satisfied.

These regulations clearly define the problems which will have to be solved by the physiology and ethology of agricultural animals. The ethologist has the task of outlining the animals' essential minimum requirements in accordance with the scientific knowledge of the day.

2. ANIMAL BEHAVIOUR

Animal behaviour and their spatial requirements vary according to species and possibly population. It is genetically determined and in addition influenced by experience made in the animate and inanimate environment.

The situation of animals kept in confinement, domesticated or not,differs from that of animals in the wild. The permanent presence of human being regarded by the animal not as dangerous but even as social partnership, the lack of natural enemies or rivals, the lack of the necessity of defending territory, the regulated feeding and last but not least the artificial selection by man (reduction of intraspecific aggression) influence animal behaviour. Apart from the endogenous factors the spatio-temporal structure i.e. the actual situation is decisive for animals' behaviour.

Different behaviour in agricultural animals compared with undomesticated animals of the same species is a result of the actual situation and therefore no indication of behavioural disturbances.

3. ETHOLOGY

Ethology of agricultural animals is an independent scientific branch which as far as possible takes into account the knowledge of basic ethologic research and of comparative ethology.

Emotional projections from the human being to the animal, which sometimes are to be found in works of K. Lorenz, have to be avoided.

In the following, I should like to point out the demands concerning modern animal husbandry systems from the ethological aspect.

I realise that one has to differentiate between desirable and absolutely necessary demands. As mentioned before, the main concern of the law is the compliance with the minimum requirements. These may in some cases lie within the optimum range.

4. MINIMUM REQUIREMENTS

4.1. Animals must at any time - undisturbed by technical installations and social partners - have the possibility to stand up or lie down in a manner characteristic to their species and to adopt a relaxed position (e.g. pigs lying on their side with stretched limbs). Space requirement can be calculated best by multiplying the height at withers by the length of the animal lying on its side (rectangle). Because of the fact that there is a high correlation between the height at withers and the length, it is sufficient to measure the height at withers and to estimate from this the length of the animal. Instead of the height at withers it is also possible to base these calculations on the weight of the animal. Space requirement then may be calculated by means of regression equation.

4.2. During the rearing period, young animals should be given enough space to satisfy their instinct of play (running, play fighting). In this respect, the structure of the stable floor is of great importance.

4.3. The stable floor must allow the animals to tread and stand upon it safely and to lie down or stand up normally without suffering injury. Perforated floors do not always fulfil these demands. Therefore at least the resting area should be littered. Dry straw is an irresistible attraction to any pig. It occupies and distracts them. Behavioural disturbances such as ear-or tail-biting and massage of the anus can normally be reduced or stopped. Littered resting areas furthermore diminish the energy input of air conditioning in the stable.

4.4. Restricted feeding requires sufficient space for each animal at the trough. Care should be taken that the animals disturb each other as little as possible.

Feeding the animals *ad libitum*, the relation animal/trough space must be selected in a way which respects the allomimetric behaviour, so that the circadian rhythm of a group is not disturbed. This can be attained if the ratio animal/trough space is 2 : 1 or at most 3 : 1.

4.5. To avoid putting individual animals at disadvantage through intraspecific aggressions (for example at feeding) animals kept in a group should be of the same weight. Ill or ailing animals must be isolated in time.

4.6. It is evident that the microclimate plays an important role in the life of a pig. Ranges of temperature, air humidity and air movement (pigs are very sensitive to wind) which can be tolerated by different age-groups, must be specified. As the pig's eye is similar to the human eye, the lighting in the barn is of great importance. The keeping of pigs in stables without windows in complete darkness should be principally forbidden.

4.7. Another important task of ethology is to indicate behavioural anomalies or disturbances caused by environmental influences and if possible to suggest means to cure them. The same is valid for technopathies, i.e. illnesses caused by technical installations in the stall.

4.8. Quantitative production performance can be no criterion as to whether the essential demands of the animals concerning locomotion, resting, comfort and social behaviour are met or not. Production yield can only signify that the quantitative and qualitative nutrient - and water supply and the microclimate are adequate, that the animal did not suffer from clinically proved illnesses which influenced production yield and that there are possible genetic differences between the animals. In the case of calves it has been shown that high yields can be achieved even if the calves are kept in very narrow boxes where they can only lie down and stand up.

On the other hand, a reduction in the production yield can be an indication that the environment created by man is not adequate.

4.9. There is no doubt that modern housing systems can improve the efficiency and quality of human labour. The large number of animals and the small amount of time required especially in automatic feeding plants may lead to disturbances in inter-relations between man-animal-environment. It can not be neglected that rationalised feeding systems, served by the human being, have the advantage that the animals' state of health can be checked by observing the appetitive behaviour.

In the case of automatic feeding plants the daily observation of each animal must be ensured in another manner.

5. APPLIED ETHOLOGY

Applied ethology is still a young scientific branch, based on basic research and dealing especially with the keeping of agricultural animals. Observation of behaviour and conclusions to be drawn from these observations are aimed at aspects concerning ecology, stable construction and thus also animal protection. Special attention has to be paid to modern housing systems, since animals kept in intensive production systems are - with respect to their most important life spheres - almost completely subjected to the environment created by human beings. Results of ethological research can directly influence development in agricultural technology if certain housing systems prove to be unsuitable for the animals. This shows the direct connection between ethology and the law for animal protection and also the great responsibility of the ethologists.

6. EXPERIMENTAL REQUIREMENT

Therefore the aim of each trial should be that it must be possible to draw conclusions applicable to the population on the basis of quantitative observations carried out on a limited number of test units. For this reason the conception of ethological trials must be based on statistical principles.

SESSION II

BEHAVIOURAL AND PHYSIOLOGICAL NEEDS OF PIGS IN GENERAL
II. PARAMETERS FOR ASSESSING WELFARE

Chairman: J.P. Signoret

CAN PHYSIOLOGICAL CRITERIA BE USED TO ASSESS WELFARE IN PIGS?

R. Dantzer and P. Mormède

Groupe INRA Stress et Physio-Pathologie de l'Environnement,
Laboratoire de Neurobiologie des Comportements,
Université Bordeaux II, 146, rue Léo Saignat,
33076 Bordeaux Cedex, France.

1. INTRODUCTION

A common approach to the assessment of welfare is the evaluation of its opposite, i.e. stress (Fraser et al., 1975; Wood-Gush et al., 1975; Dantzer and Mormède, 1979; Syme and Syme, 1979). This is particularly the case when physiological criteria are taken into consideration, since there is no direct physiological evidence of welfare. In contrast, it has long been known that exposure to physical or psychological stressors elicits a wide range of physiological changes in the organism which can be more or less easily detected.

Two main reactions to stress have been identified:

- the emergency reaction shown by Cannon (1911) to be related to the activation of the sympathetic nervous system and the adrenal medulla. This is a short latency response involving hormonal factors (catecholamines) which enable the subject to mobilise its resources quickly for the metabolic requirements of fighting or flight;
- the general adaptation syndrome, initially described by Selye (1950) and characterised by the release from the anterior pituitary of adrenocorticotrophic hormone (ACTH) which in turn activates the release of corticosteroids from the adrenal cortex. Corticosteroids amplify and further extend the metabolic and vascular effects of catecholamines.

Evidence for the occurrence of any of these reactions is based either on direct measurement of the hormonal levels in the blood or on the evaluation of secondary physiological or metabolic changes elicited by the released hormones.

For historical and technical reasons, most research efforts have concentrated on the pituitary-adrenal system. Many stimuli or conditions have been shown to elicit an increase of plasma corticosteroids. This change has therefore been proposed as the operational definition of stress, i.e. animals which exhibit higher than normal plasma corticosteroid levels are claimed to be in a state of stress and stimuli or conditions which elicit such hormonal changes are described as stressors. However, in intensive husbandry, the term stress is used more loosely to describe any set of conditions leading to an unexplained drop in production or to pathological conditions (e.g. transport, weaning, social stress, etc.). In addition, most studies on stress in pigs have concentrated on the so-called porcine stress syndrome, a lethal pathological condition triggered by environmental events, which was initially believed to be due to a failure of adaptative mechanisms[1].

In this article, I do not want to enter endless controversies about the many meanings and uses of the term stress. I shall restrict myself to a prospective evaluation of the meaning of hormonal reactions to environmental stimulation, with the hope that such a knowledge will help to determine what are the interest and the pitfalls of the use of plasma hormonal levels to assess stress in pigs. For that purpose I shall consider the pituitary-adrenal axis which is the only neuro-endocrine system for which there are enough data at the present time to achieve such an analysis.

2. COGNITIVE ASPECTS OF PITUITARY-ADRENAL ACTIVATION AND NON-SPECIFICITY IN STRESS THEORY

Pituitary-adrenal activation is demonstrated by higher than normal plasma ACTH or corticosteroid levels. This reaction

[1] Recent data unequivocally indicate that the porcine stress syndrome or, more exactly, its model, the malignant hyperthermia syndrome elicited by halothane, has no behavioural nor neuro-endocrine correlates (Dantzer and Mormède, 1978; Mormède and Dantzer, 1978; Dantzer and Hatey, 1981) but is specifically related to a defect of the muscle mitochondria in fixing calcium ions (Cheah and Cheah, 1976).

can be observed under a wide range of environmental conditions (e.g. handling, climatic conditions, casualties, etc.) and is therefore said to be non-specific.

As can be seen from Figure 1, psychological stressors (exposure to novelty) are as effective as physical stressors (electric shock) in increasing plasma corticosteroids. Moreover in an aggressive situation in which 2 pigs from different social groups are put together, the increase in plasma corticosteroids depends on the presence or absence of fighting and is more pronounced in subordinates than in dominant animals.

These data indicate that the pituitary-adrenal system is sensitive to the psychological components of the stress situation. It can be shown that this sensitivity to psychological factors is responsible for the apparent non-specificity of pituitary-adrenal activation (Figure 3). This conclusion is somewhat different from the one which was initially formulated by Selye (1950), for whom non-specificity was related to the consequences of pituitary-adrenal activation: whatever its physical dimension, a stressor was seen as disturbing homeostasis and therefore triggering physiological mechanisms aimed at reinstating in a non-specific way the initial equilibrium. In the present analysis, the physiological response to stress is non-specific not because of the response but because of the psychological components contained in any environmental stimulus presented to a conscious subject. This idea has 2 main consequences:

1. Stress reactions are controlled at a higher level of integration than a simple reflex (i.e. limbic system and cortical areas) (Henry and Stephens, 1977).
2. The psychological characteristics of exteroceptive stimuli can be further subdivided into operational components such as novelty, conflict, uncertainty and active adjustment possibilities (coping), each of which can eventually be studied on its own (Hennessy and Levine, 1979).

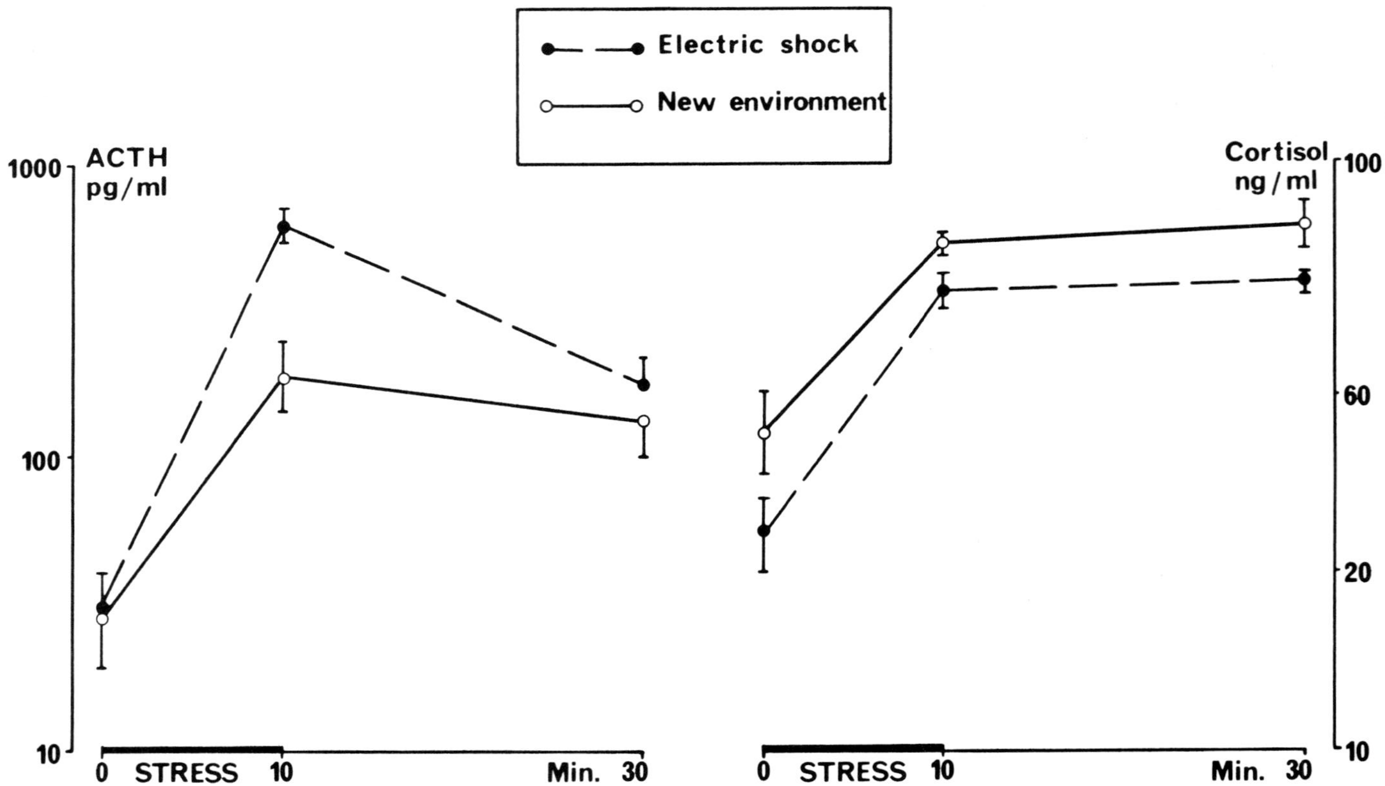

Fig. 1. Pituitary-adrenal response of pigs to inescapable electric shocks and to a novel environment. Eight pigs were submitted to a 10 min exposure to a novel environment and, 1 week later, to a 10 min session during which 15 brief inescapable electric shocks (25 mA) were delivered. ACTH was measured by radioimmunoassay and circulating corticosteroids by a competitive protein binding method in plasma taken by puncture of the *vena cava* at time 0, 10 and 30 min. Each value is the mean of 8 determinations (Mean ± SEM).

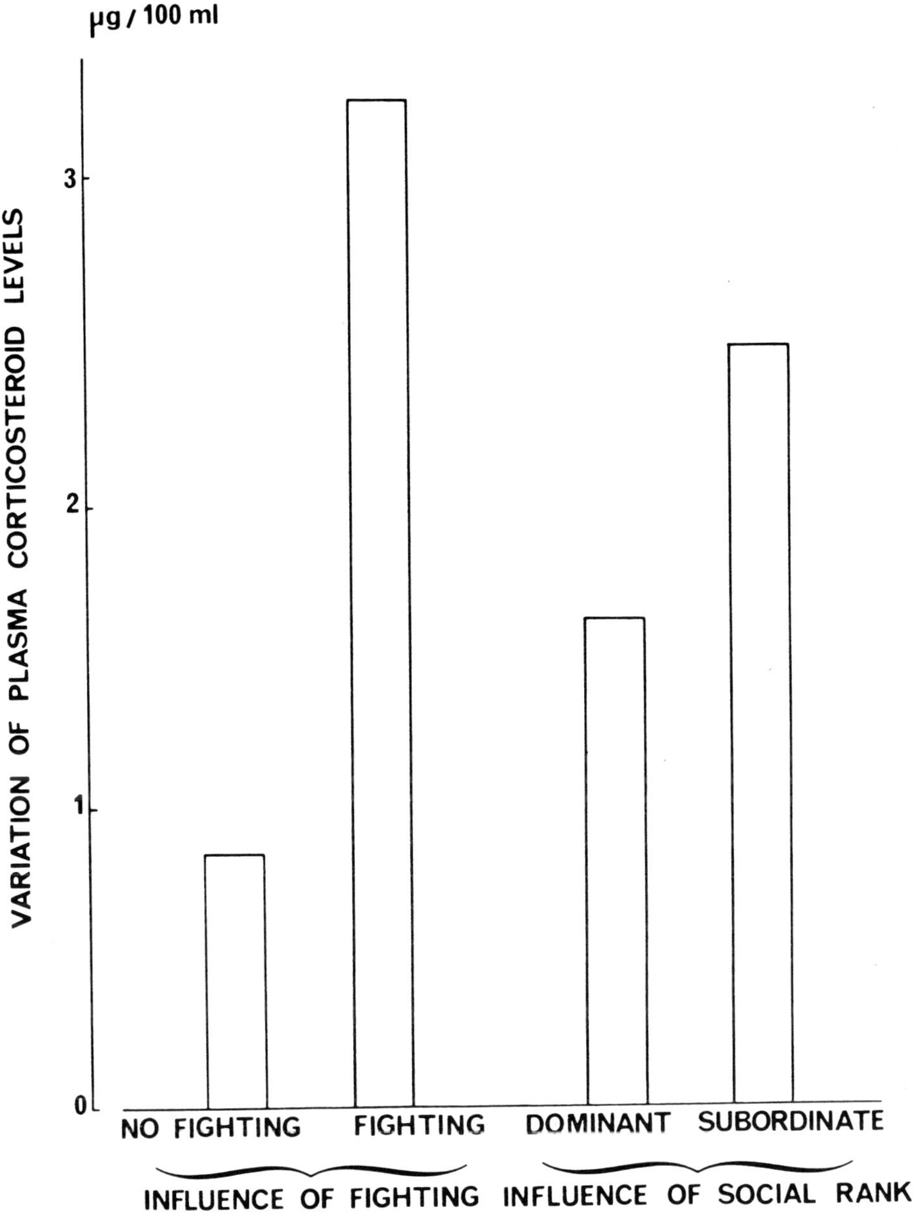

Fig. 2. Pituitary-adrenal response of pigs to social factors. When 2 strange pigs are put together they engage in aggressive behaviour. Variation of circulating corticosteroids (i.e. postsession plasma corticosteroid levels minus presession corticosteroid levels) is more pronounced when the animals fight than when there are only pushing and biting episodes. In both cases, however, the increase is more pronounced in subordinates than in dominant animals. Each value is based on the observation of 12 pigs and differences between means are highly significant (Arnone and Dantzer, 1980).

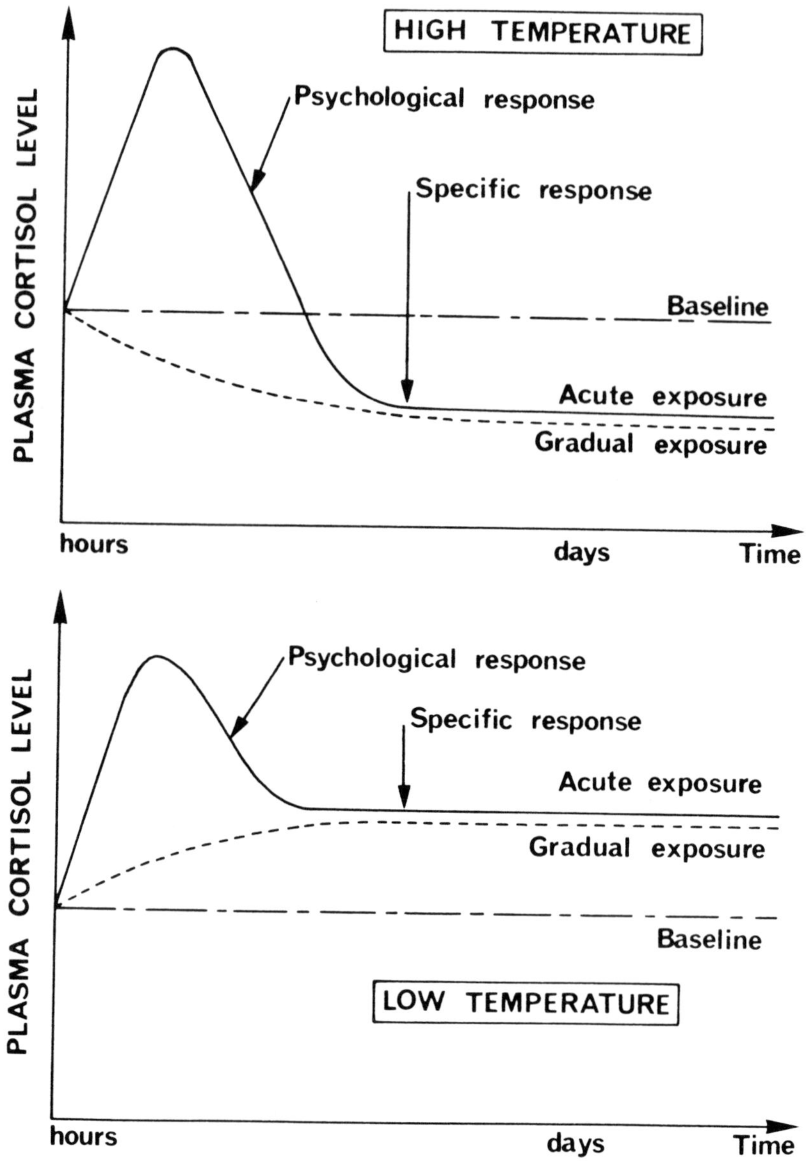

Fig. 3. Diagram of pituitary-adrenal response of animals to thermal stress. If animals are acutely exposed to either warm or cold temperature, circulating cortisol levels increase and then decrease before stabilising at a lower level (warm exposure) or at a higher level (cold exposure) than normal. If great care is taken to accustom the subject to handling and to the environment of the metabolism cage and if the temperature is gradually changed, the abrupt variation of plasma cortisol levels disappears and is replaced by a gradual change to the final levels. The abrupt variation of plasma cortisol levels reflects the emotional reaction to the whole situation while the gradual change specifically reflects the reaction to the physical quality (i.e. warm or cold) of the environmental temperature (Dantzer and Mormède, 1979).

3. HORMONE-BEHAVIOUR RELATIONSHIPS IN AVERSIVE SITUATIONS

Hormonal reactions to stress do not occur in an inanimate organism but in a behaving subject who may adopt one of several attitudes to cope with the situation. It is therefore of interest to study the relationships between hormonal reactions and behavioural responses to the situation. These relationships can assume 2 modes, i.e. hormones can affect behavioural responses and, conversely, endocrine status can be modified if the subject has certain experiences and engages in specific behaviours (Leshner, 1978).

3.1 Effect of hormonal activity on behavioural responses

Two types of evidence are available for the demonstration of the existence of hormonal effects on behaviour.

Correlational evidence is based on the relationship of circulating corticosteroid levels with the form and intensity of particular classes of behavioural responses. For instance, it has been shown that the amount of response emitted on extinction of a food reinforced operant is inversely related to the baseline plasma corticosteroid levels, i.e. pigs which have learnt to get food by pushing a panel with their snout stop responding when their response is no longer rewarded (extinction). Pigs with higher baseline circulating corticosteroid levels stop earlier than pigs with lower circulating corticosteroid levels (Figure 4).

Experimental studies allow the determination of the causality in correlational relationships by manipulating the hormonal state and looking at the effect of that manipulation on the individual's reaction to exciting stimuli. For instance, pigs which have learnt to avoid unsignalled electric shock by jumping from one compartment to the other in a shuttle-box, react to a fear signal (i.e. a tone previously paired with inescapable electric shock) by increasing their response rate. Such a reaction is exaggerated by increasing circulating corticosteroid levels

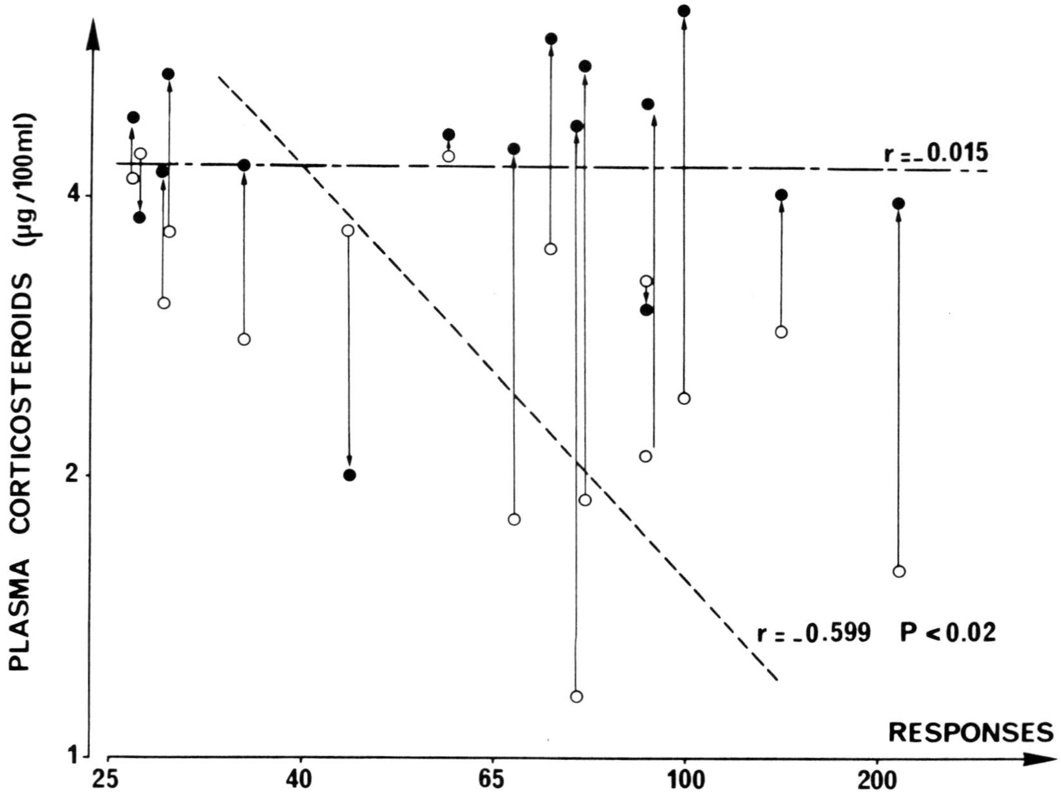

Fig. 4. Relationship between the number of panel presses emitted during an extinction session in pigs which have learnt to get a food reward by pressing the panel with their snout and the plasma corticostercid levels. Open circle shows for each pig the initial plasma corticosteroid levels measured just before the extinction session, and closed circle the plasma corticosteroid levels measured 30 min after the completion of the extinction session. Response rates were correlated to initial plasma corticosteroid levels (r = -0.599) but not to final plasma corticosteroid levels (r = -0.015) (Dantzer et al., 1980).

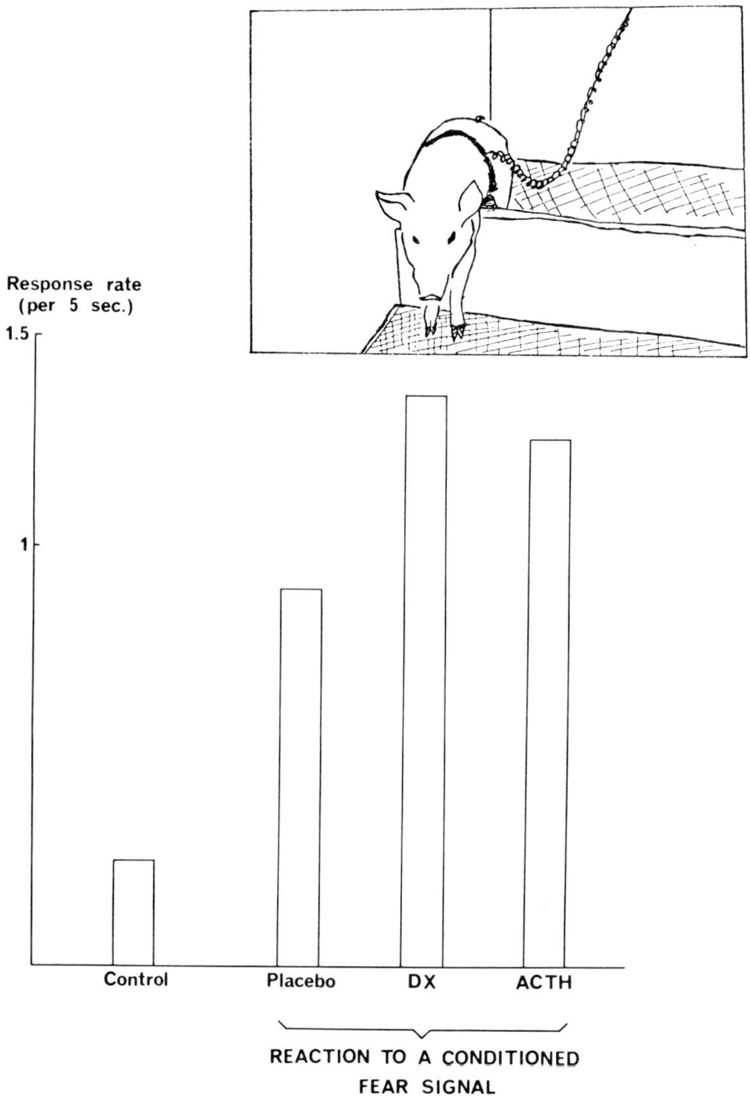

Fig. 5. Pituitary-adrenal influences on reaction to a fear signal in pigs. Pigs trained to avoid electric shocks by jumping from one compartment to the other in shuttle-box (upper diagram) are confronted with a fear signal, i.e. a tone which has previously been paired with inescapable electric shocks. The presentation of the fear signal induces an increase of response rate from 0.25 response/5 sec to about 0.9 response/5 sec. The previous injection of 0.2 mg/kg dexamethasone or 6.25 IU/kg ACTH1-24 significantly increases the the reaction to the fear signal (Mormède and Dantzer, 1978).

either by ACTH injection or by supplying exogenous corticoids (Figure 5).

3.2 Effects of behavioural attitudes on hormonal activity

The kind of behaviour the subject engages in may be critical for the nature and intensity of hormonal responses. This is particularly clear when comparing animals differing by their ability to adjust in an active way to the stressful situation. For instance, the high levels of plasma corticosteroids observed in pigs submitted to a few inescapable electric shocks contrast with the low levels observed in pigs which have learnt actively to avoid shocks but which still receive the same amount of electric shock as the former group.

It can also be shown that when animals are submitted to conflict situations, engaging in activities apparently unrelated to the situation (displacement activities according to ethologists, adjunctive behaviour according to psychologists) (Falk, 1971) has a profound influence on pituitary-adrenal hormone levels. For instance, hungry pigs submitted to intermittent food deliveries develop stereotyped chain nibbling in the interval between food deliveries (Figures 6 and 7). By comparing presession and postsession plasma corticosteroid levels, it can be seen that this apparently abnormal behaviour has homeostatic properties, i.e. it enables the subject to lower its circulating corticosteroid levels (Figure 8). It can therefore be seen as a very effective means by which the animal can dissipate its tension or anxiety.

4. FACTORS INFLUENCING HORMONE-BEHAVIOUR RELATIONSHIPS

Hormone-behaviour relationships can be modified either by genetic factors or by previous experience. Evidence available in pigs suggests that genetic factors are more important than previous experience.

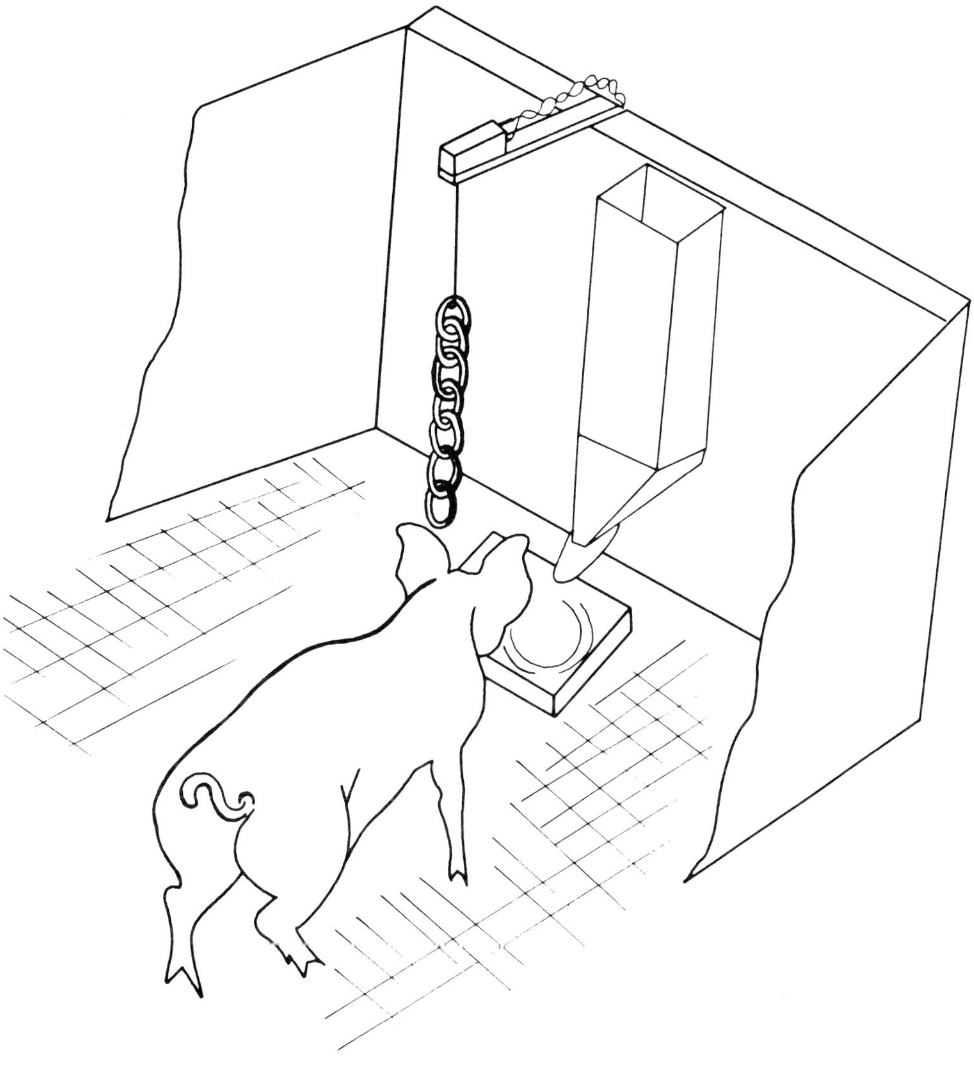

Fig. 6. Experimental procedure to study displacement activities in pigs. Hungry pigs are submitted to intermittent food deliveries (20 g each 5 min). When a chain hanging from the ceiling is simultaneously available, pigs develop stereotyped chain nibbling in the interval between food deliveries. This behaviour can be recorded by the operation of a microswitch to which the chain is connected.

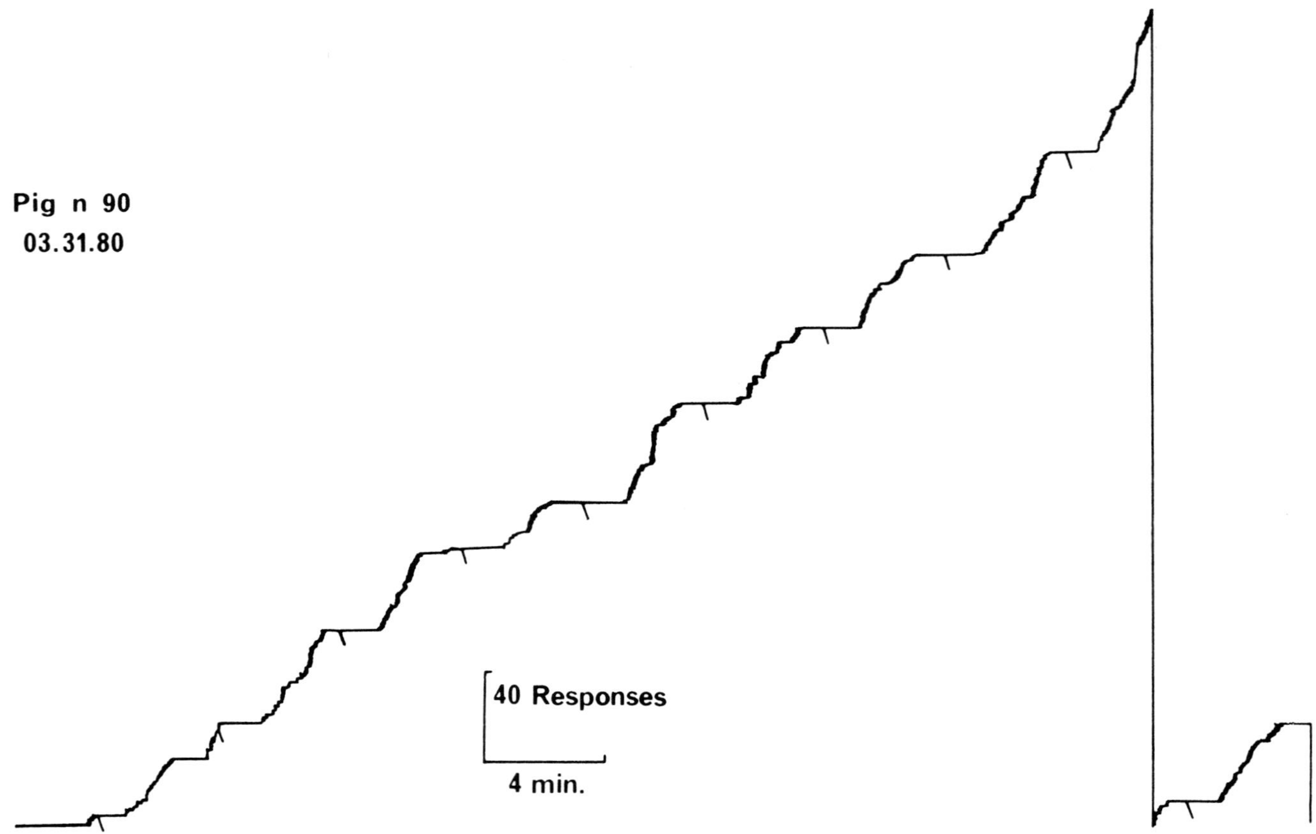

Fig. 7. Cumulative graph illustrating the typical behaviour of pigs submitted to intermittent food deliveries. A downwards deflection of the pen occurs each time food is delivered. An upwards deflection of the pen occurs each time the animal nibbles at the chain. Note that the pig starts chain nibbling after eating the food received and stops before the next food delivery.

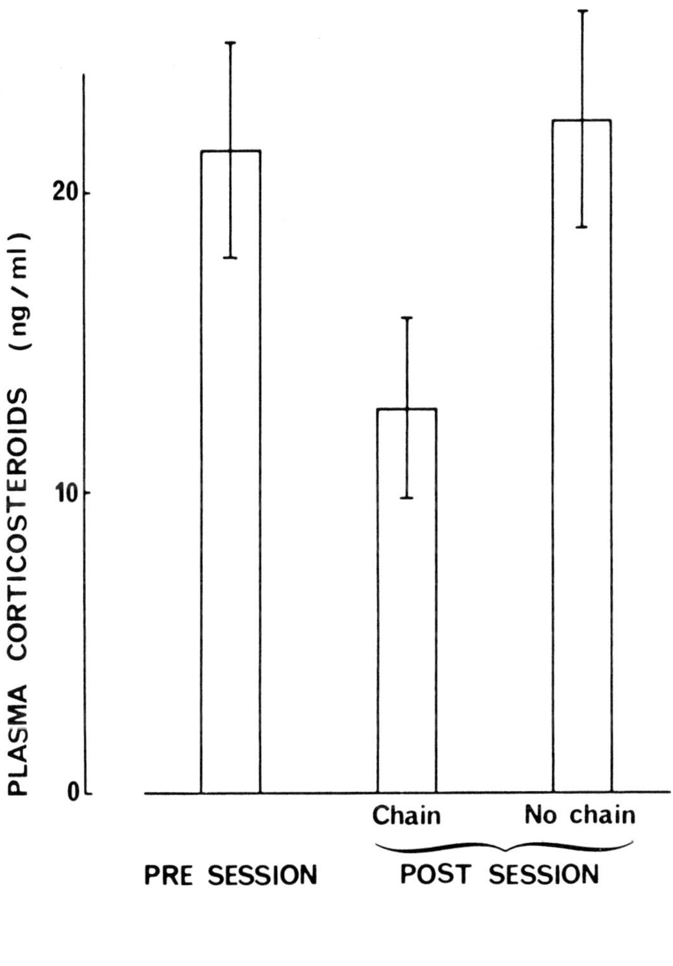

Fig. 8. Circulating corticosteroid levels (mean ± SEM, n = 8) in pigs showing stereotyped chain nibbling. Postsession levels are significantly lower than presession levels, and this variation disappears when the chain is no more available.

4.1 Differences between populations

Apart from sensitivity to the porcine stress syndrome, little systematic study has been conducted on the behavioural and neuro-endocrine characteristics of the different breeds of pigs used in intensive husbandry. In a study of 2 different populations of pigs, Piétrain pigs and Large White pigs, kept in the same husbandry conditions, it has been shown, by combining various behavioural tests and measurements of pituitary-adrenal activity, that Large White pigs are characterised by a relative inability to initiate active response when confronted with aversive events, this inability being correlated to higher baseline pituitary-adrenal activity (Dantzer and Mormède, 1978; Mormède and Dantzer, 1978). More research is required to find out if such differences can be taken as a basis for the selection of strains of animals that are more suited to intensive conditions or have better adaptive abilities.

4.2 Influence of experiential factors

Many studies on rodents and on primates have demonstrated the influential role played by early experience on subsequent adaptive abilities of animals (Denenberg, 1969). In contrast to the large bulk of data concerning behaviour, little information is available on the long-term effects of either husbandry techniques or chronic stress conditions on pituitary-adrenal activity. We have observed that either social stress (i.e. mixing 2 groups of animals from different origins 3 days before the test and inducing competition for feeding at the trough) or avoidance conditioning has no effect on circulating corticosteroid levels nor on pituitary-adrenal reactivity to a standardised stressor (Figures 9 and 10). In the same manner, plasma ACTH levels of piglets weaned at 3 weeks and kept in flat-deck did not differ from those of piglets weaned at 6 weeks in the maternity pen, when the animals were submitted to a novel environment at 7 weeks of age (Figure 11).

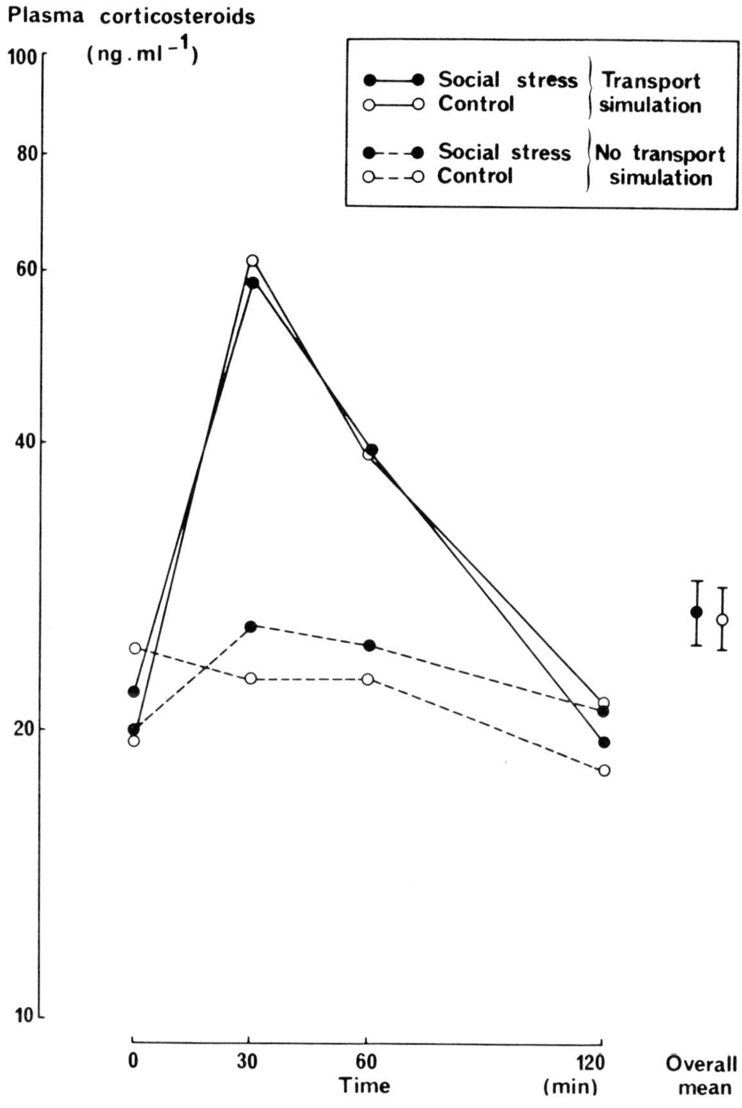

Fig. 9. Pituitary-adrenal response, as measured by circulating corticosteroid levels (mean ± SEM, n = 4), to a standardised stressor, of control pigs and of pigs submitted to social stress and crowding from 3 days before the test. Each group contained 8 pigs and was subdivided into 2 groups, 1 group submitted to shaking during 25 min and 1 group remaining in its pen. Blood samples were taken at time 0, 30, 60 and 120 min. The only significant variations were due to the shaking stress.

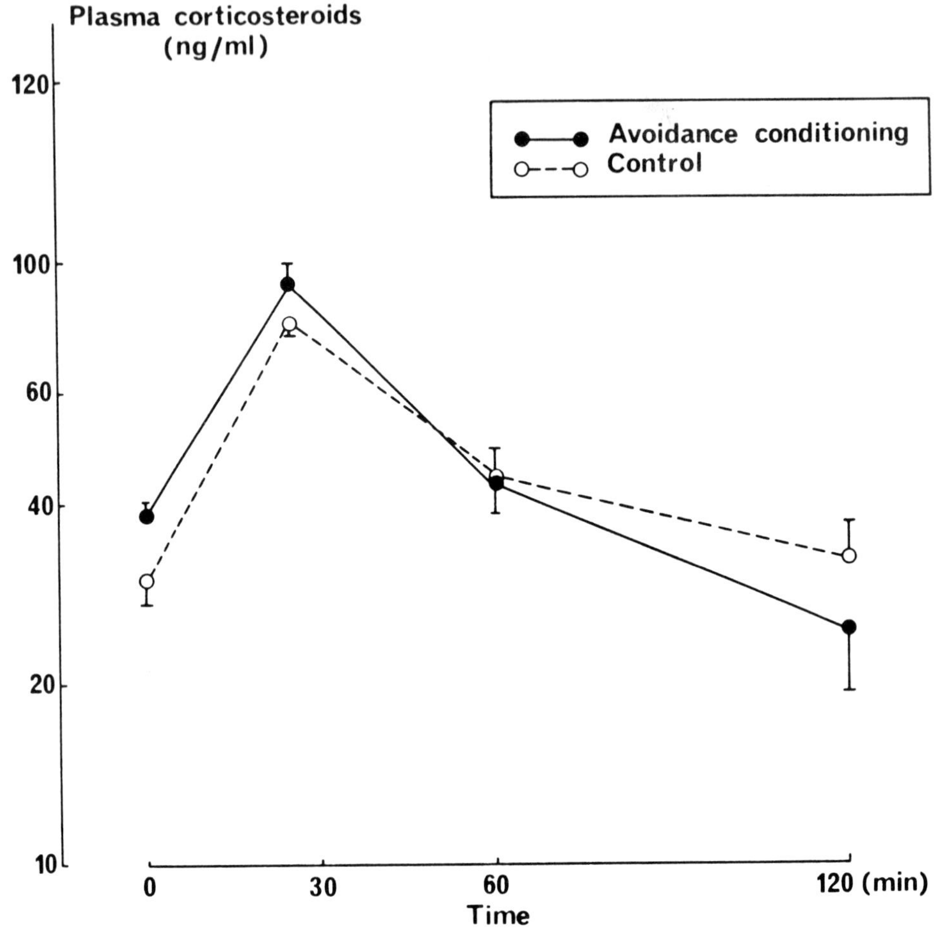

Fig. 10. Pituitary-adrenal response, as measured by circulating corticosteroid levels (mean ± SEM, n = 4), to a standardised stressor, of control pigs and pigs submitted to avoidance conditioning from 2 weeks before the test. Each group of 4 pigs was submitted to shaking for 25 min. Blood samples were taken at time 0, 30, 60 and 120 min. The only significant variations were due to the shaking stress.

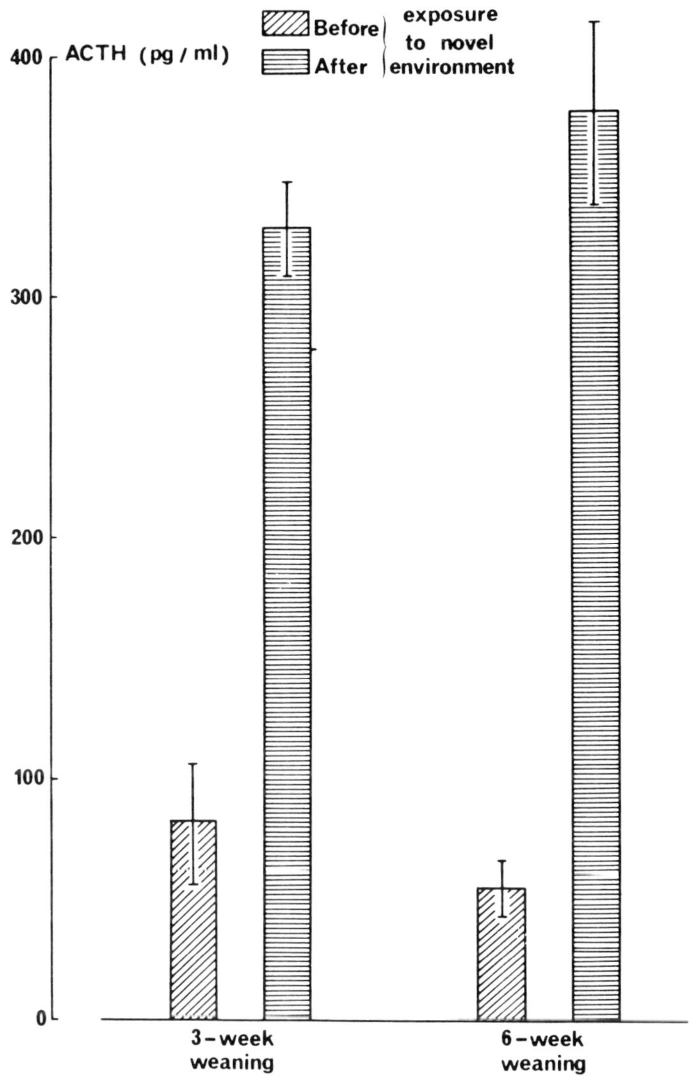

Fig. 11. Plasma ACTH levels (mean ± SEM, n = 8) of 7 week old piglets exposed to a novel environment during a 15 min session. Blood was withdrawn before the session (diagonally hatched columns) and at the end of the session (horizontally hatched columns). Piglets weaned at 3 weeks and kept in flat-deck did not significantly differ from piglets weaned at 6 weeks in the maternity pen.

5. CONCLUSION

I. In the pig, the pituitary-adrenal axis is very sensitive to short-term stressors, in which case, when tested in conscious animals, its activation reflects mainly the psychological aspects of the stress situation (i.e. its degreee of novelty or the ability of the subject to cope with the situation, etc.).

II. Behavioural responses are profoundly influenced by hormonal activity, which means that pituitary-adrenal hormones do affect the state of the central nervous system circuits involved in the perception of the stressful stimuli and the integration of the relevant responses. In addition, the behaviour of the subject produces changes in its hormonal state, probably through the effects of these experiences and actions on neural activity. All these factors make obsolete the original view of stress reaction as a reflex and underline the need to consider both behavioural and hormonal responses in an integrated and dynamic way.

III. Hormone-behaviour relationships can be influenced by genetic factors, although their exact role has still to be determined. In contrast, experiential factors appear to have little influence on pituitary-adrenal baseline activity and responsiveness. This means that circulating ACTH and corticosteroid levels are not sensitive to chronic stress. This result can be taken in 2 ways:

1) either as demonstrating the good adaptive abilities of domestic pigs, or

2) as an indication of the limitations of classical endocrinological measurements.

In this case, other parameters need to be studied, for instance, nocturnal rhythms, CBG affinity and binding capacity, receptors, etc.). However, it must at the same time be borne in mind that the pituitary-adrenal system is not the sole system activated in stress situation and that other neuro-hormonal systems (e.g. catecholamines, growth hormone, thyroid hormones, insulin and glucagon, endorphins, etc.) certainly play a very important role.

IV. Research efforts are urgently needed to define the role of these systems and their participation in stress reactions.

REFERENCES

Arnone, M. and Dantzer, R., 1980. Does frustration induce aggression in pigs? Appl. Anim. Ethol., 6: 351-362.

Cannon, W.B. and de la Paz, D., 1911. Emotional stimulation of the adrenal gland. Amer. J. Physiol., 27: 64-70.

Cheah, K.S. and Cheah, A.M., 1976. The trigger for PSE condition in stress-susceptible pigs. J. Sci. Fd. Agric., 27: 1137-1144.

Dantzer, R. and Mormède, P., 1978. Behavioural and pituitary-adrenal characteristics of pigs differing by their susceptibility to the malignant hyperthermian syndrome induced by halothane anesthesia. I. Behavioural measures. Ann. Rech. Vet., 9: 559-567.

Dantzer, R. and Mormède, P., 1979. Le stress en élevage intensif. Masson, Paris.

Dantzer, R. and Hatey, F., 1981. Plasma dopamine-beta-hydroxylase and platelet monoamine oxidase activities in pigs with different susceptibilities to the malignant hyperthermia syndrome induced by halothane. Reprod. Nutr. Develop. (in press).

Dantzer, R., Arnone, M. and Mormède, P., 1980. Effects of frustration on behaviour and plasma corticosteroid levels in pigs. Physiol. Behav., 24: 1-4.

Denenberg, V.H., 1969. The effects of early experience. In: Hafez E.S.E. (ed.). The behaviour of domestic animals. Baillière, Tindall and Cassell, London, 2nd edition, pp. 95-130.

Falk, J.L., 1971. The nature and determinants of adjunctive behaviour. Physiol. Behav., 6: 577-588.

Fraser, D., Ritchie, J.S.D. and Fraser, A.F., 1975. The term stress in a veterinary context. Brit. Vet. J., 131: 653-662.

Hennessy, J.W. and Levine, S., 1979. Stress, arousal and the pituitary-adrenal system: A psychoendocrine hypothesis. Progr. Psychobiol. Physiol. Psychol., 8: 133-178.

Henry, J.P. and Stephens, P.M., 1977. Stress, health and the social environment. A sociobiological approach to medicine. Springer Verlag, New York.

Leshner, A.I., 1978. An introduction to behavioural endocrinology. Oxford University Press, New York.

Mormède, P. and Dantzer, R., 1978. Behavioural and pituitary-adrenal characteristics of pigs differing by their susceptibility to the malignant hyperthermia syndrome induced by halothane anesthesia. II. Pituitary-adrenal function. Ann. Rech. Vet., 9: 569-576.

Mormède, P. and Dantzer, R., 1978. Pituitary-adrenal influences on avoidance behaviour in pigs. Horm. Behav., 10: 285-297.

Selye, H., 1950. Stress. Acta Inc., Med. Publ., Montreal.

Syme, G.J. and Syme, L.A., 1979. Social structure in farm animals. Develop. Anim. Vet. Sci., 4: 1-200.

Wood-Gush, D.G.M., Duncan, I.J.H. and Fraser, D., 1975. Social stress and welfare problems in agricultural animals. In: Hafez, E.S.E. (ed.). The behaviour of domestic animals, Baillière, Tindall and Cassell, London, 3rd edition, pp. 182-200.

DISCUSSION

V. Tarrant *(Ireland)*

Dr. Dantzer, will you please explain, in terms of behaviour, what factors you think determine whether or not strange pigs will fight or show aggressive behaviour when they are brought together? I refer particularly to your Figure 2.

R. Dantzer *(France)*

There have been some studies on the factors involved in perception, whether it is visual, olfactory, and so on. However, it is very complex since individual recognition in pigs is not unique. You can have individual recognition between pigs from different social groups, and from the same group, but it is not the same thing. However, I am not an expert in this field.

V. Tarrant

In your comments on halothane positive and halothane negative Piétrain pigs, you observed very little difference between those two groups. Were the halothane negative pigs heterozygous or homozygous with respect to the halothane gene.

R. Dantzer

This has not been checked. This work was done with the genetic department at our Institute. As you know, the positives are always homozygous but the negatives can be either heterozygous or homozygous. We don't know the exact state of the halothane negative group. What is important is that these pigs were from the same litter and they were reared in the same conditions, so there were no differences in terms of early experience. The study has shown that there are no behavioural differences between these pigs. They react in exactly the same way to different situations, for example, an open field test, avoidance conditioning, active avoidance, passive avoidance, etc. However, there are wide breed differences, Large White are not the same as Piétrain.

PARAMETERS FOR ASSESSING WELFARE, ETHOLOGICAL CRITERIA

Doris Buchenauer
Institut für Tierzucht und Tierfütterung der Universität Bonn, Federal Republic of Germany

1. INTRODUCTION

The introduction of intensive husbandry systems in pig production has resulted particularly in changes of feeding methods and removal of animal manure.

Whereas, formerly, the animals were occupied for quite long periods in consuming bulk feeds, nowadays concentrates are given which require only a short time for consumption.

The system of keeping the animals indoors throughout the year has been substituted for free range systems. Slatted, or other perforated floors, are used increasingly instead of solid floors with straw litter in all areas of pig production. These changes are related to an increase in stock density which, at the same time, means a decrease of space per animal and, because of the barren environment, have resulted in difficulties of adaptability and have led to welfare problems.

Welfare is defined by the Brambell Report (1965) as a wide term that embraces both the physical and mental well-being of the animals. The Dutch Welfare Report (1977) based on the Lorz (1973) definition, stated that welfare is the existence in reasonable harmony within the animal itself and with the environment, both from an ethological and a physiological point of view. Kilgour (1980) noted that a shift in emphasis to the animal's behaviour rather than its mental well-being is a feature of this definition.

The German law for the protection of animals (1972) requires protection of life and well-being. Well-being is defined by the German legislators as an undisturbed behaviour pattern, appropriate to the species, and corresponding to a series of biological phenomena.

All the definitions mentioned above may be applied only to those animals whose habitat does not exceed the limits of their adaptability. According to Tschanz (1978) adaptation is achieved on the one hand by the evolving of features which are especially suitable for a given environment and, on the other hand, by changes of existing features in the direction of improved efficiency. The same author bases the adaptation value (as an expression of the adaptability to specific living conditions) on the reproductive capability of a genotype. This criterion includes the number of surviving offspring regardless of whether this is attributable to enhanced longevity, higher sexual activity or increased fertility. According to Wood-Gush et al. (1975), agricultural animals which for thousands of generations have been kept under extensive or semi-extensive conditions are likely to be frustrated, thwarted and in conflict under modern husbandry conditions.

In applied animal ethology, according to Duncan (1980), there are three methods available for assessing welfare by behaviour:

1. By looking for unusual or inappropriate behavioural changes.
2. By giving the animals a choice of environments and assuming that they will choose in the best interests of their own welfare.
3. By subjecting the animals experimentally to stress situations, observing their behaviour and comparing it to that which occurs under commercial conditions.

Other authors recommend measuring behaviour by studying the animals in an environment which is assumed to be ideal, or which can at least be used as a reference system, and then comparing behaviour with that of animals in the environment under investigation (Hughes, 1976; van Putten, 1977). This procedure is necessary since well-being cannot immediately be expressed by the animal. Smidt et al. (1980) defined well-being as being a reflection of a state of relative psychological and physical integrity of an individual which can only be ascertained by drawing conclusions from certain symptoms, for example, the behavioural reactions.

2. KINDS OF BEHAVIOUR

In the following I would like to focus on three categories of behaviour which may be taken into consideration in assessing welfare:

1. Occurrence of pain and fear.
2. Abnormalities in behaviour.
3. Modifications in behaviour.

2.1 Occurrence of pain and fear

A state of well-being in pigs must exclude suffering from pain and fear. The expression of these states has been defined by Grauvogl (1972) (Table 1). According to Grauvogl (1978), the term pain in the ethological sense covers the total expressions by the animal of disturbed well-being - the disturbed course of measurable physical functions. The same author defines fear, in the sense of welfare, as the animal being exposed to acute environmental stress leading to conflict situations which cannot be counteracted by the behaviour patterns of the animal, and which may thus lead to a change in the situation, giving rise to enormous tension which justifies the assumption that one may consider fear as a special form of pain. These means of expression can be detected visually, by fast or slow photography, by telemetry equipment and sound spectrographs. The frequencies of the data obtained can be statistically evaluated.

TABLE 1

EXPRESSION OF PAIN AND FEAR BY PIGS

Expression of pain	Expression of fear
1. Shrieking	1. Increased frequency of defaecating and urinating
2. Inaudible groaning	2. Trickle of urine (caused by fright)
3. Clenched jaws, grinding of teeth	3. Pale skin, bristling hair, trembling body
4. Increased blood circulation and breathing rate	4. Increased frequency of heart beat and increased breathing rate
5. Curvature of body	5. Changing colour
6. Disturbed movements	6. Dilated eyes, nose and mouth
7. Thrusting out movement of painful legs	7. Moderately loud vocalisation
8. Animals lying close together to combat feeling cold	8. Unnatural huddling together and climbing on other animals
9. Drowsiness	9. Displacement activity (suckling another animal, massage of the udder)
10. Motionless eyes and stiff ears	10. Alienated behaviour, e.g. panic flight in the sense of the animal 'losing its head'.

2.2 Abnormalities of behaviour

A further ethological criterion for assessing welfare is the occurrence of disturbed behaviour. All behavioural patterns which differ from the norm for a species can be categorised as disturbed behaviour. The abnormalities of behaviour observed in pigs are listed in Table 2.

TABLE 2

ABNORMALITIES IN THE BEHAVIOUR OF PIGS

Type of abnormality	Suckling piglets	Early weaned piglets	Fattening pigs	Sows
Tail and ear biting	x	x	x	
Cannibalism		x	x	
Belly nibbling		x		
Tongue rolling		x	x	x
Rubbing nasal bone		x	x	x
Rubbing snout		x	x	x
Hyperactivity		x	x	x
Massage of anus			x	
Vacuum chewing				x
Biting the bars of the pen				x

Depending on modality, intensity or frequency, these abnormalities may occur as different redirected activities, for instance, as stereotyped behaviour or as vacuum activities. Pigs kept in intensive stable systems exhibit disturbed behaviour in all stages of development. Tail and ear biting is observed as early as 3 to 4 weeks of age in suckling piglets kept on perforated floors without straw. Early weaned piglets often suck at the navels, ears and preputium, and massage the bellies of their pen mates, because of the unsatisfied need to suckle. Achebe (1975) noted that piglets weaned at the age of 3 weeks were still sucking each other in the eighth week of life. Early weaned piglets kept in flat decks show unnatural behaviour by rapid changes of motor activity which gives the impression of extreme restlessness. Such activity is often related to abnormalities such as tongue rolling and flicking, rubbing of the nasal bone and snout. This behaviour cannot be prevented in stables with dim lighting or in the dark (Schrenk, 1978). Tail and ear biting also occurs with early and later weaned piglets which are kept in a poor environment.

Tail biting is one of the most frequently occurring behavioural disturbances of fattening pigs. Its causes have been described in detail by van Putten (1968, 1977, 1978a). The increase of these disturbed behavioural patterns from year to year can be equated with the increasing use of intensive housing systems (van Putten, 1978a). Lindquist (1977) found that tail biting occurred in 50% of 155 farms investigated; among these only a few cases were recorded in housing systems on straw.

From investigations in Dutch slaughter houses it appears that tail docking as a precautionary measure could reduce the frequency of tail biting to some extent, but could not prevent it (Meijer et al., 1976). However, Haske-Cornelius (1977) found that tail docking did not affect the extent of the occurrence of either tail or ear biting. Anal massage related to coprophagy has been observed by Sambraus (1978).

The type of pigs most affected by the extreme changes in housing systems are sows. In many cases their adaptability has been obviously exceeded since whole stocks have reacted with the abnormal patterns of behaviour described in Table 2. According to Sambraus (1980) on two farms, 17.5% of the behaviour patterns observed consisted of abnormalities when pregnant sows were kept in intensive housing systems, whereas in the same stable systems, using sawdust and supplying roughage, only 7.1% of the behaviour patterns were abnormalities. In some cases where lack of exogenous stimuli was the cause of abnormalities, even minor changes helped the animals: for example, small amounts of litter or hay feeding proved to be effective.

The occurrence, and frequently the high rate of occurrence, of disturbed behaviour as described in Table 2, is considered to be an indication of frustration and therefore of the undesirability of the housing system used.

2.3 Modifications in behaviour

A further method for assessment of welfare includes investigations of typical patterns of the appropriate behaviour as ethological parameters. Reference systems are compared with intensive housing systems. Table 3 shows the characteristic behavioural patterns of pigs.

TABLE 3

CHARACTERISTIC BEHAVIOUR PATTERNS OF PIGS

Suckling pigs	Weaned pigs	Fattening pigs	Sows
Sucking	Eating	Eating	Eating
Drinking	Drinking	Drinking	Drinking
Eating	Rooting	Rooting	Rooting
Rooting	Playing	Exploring	Exploring
Playing	Exploring	Playing	Walking
Exploring	Agonistic behaviour	Agonistic behaviour	Mating behaviour
Agonistic behaviour	Comfort behaviour	Comfort behaviour	Nesting behaviour
Comfort behaviour	Eliminative behaviour	Walking	Maternal behaviour
Eliminative behaviour	Walking	Eliminative behaviour	Agonistic behaviour
Walking	Resting	Resting	Eliminative behaviour
Resting			Resting

These patterns of behaviour change under poor environmental conditions. On the basis of the literature and from my own observations comparing standard and intensive stable systems, the changes of behavioural patterns have been classified in terms of decreased and increased occurrence.

These evaluations show that young animals react predominantly by decreased behavioural activities appropriate to the species (Tables 4 and 5). Rooting, playing, walking, investigating and exploring are especially affected. Playing in particular may be regarded as a sensitive indicator for assessing welfare since it occurs only in the absence of basic needs. Playing is of essential importance for the normal development of an animal; it provides the possibility for a dialogue between the animal and its environment (Meyer-Holzapfel, 1956; Eibl-Eibesfeld, 1972). Rooting can be considered as a means for the animal to explore its environment. Weiss (1975) concludes that a diverse structured environment which stimulates exploration and investigation or playing with objects, results in an improved adaptability of the animal.

Important behaviour patterns for young animals are summarised by Hassenstein (1974) as follows: during exploratory behaviour the animal becomes acquainted with a definite spatial environment and with investigation behaviour it becomes acquainted with unfamiliar objects; during playing it develops its own ability of action by interplaying with the environment. Intensive stable systems can cause deprivations which prevent the satisfaction of natural physical and emotional needs in consummatory acts. This may lead to behavioural modifications and abnormalities. Hypertrophic behaviour patterns such as massaging, sucking and nibbling on pen mates, as well as nibbling on objects, show that the physical and emotional need to suck is not fulfilled.

Deplaced and frequent eliminations, as well as timidity and huddling together are indications of fear (Table 4).

Frenzied activity corresponds with the increase of the intensity of preceding behaviour which Hinde (1973) regards as an indication of frustration. The higher frequencies of sitting and standing may partly be considered as conflict behaviour.

Table 6 presents the reactions which appear in the

TABLE 4

MODIFICATIONS OF BEHAVIOUR OF EARLY WEANED PIGLETS IN A BARREN ENVIRONMENT

Behaviour pattern Hypotrophic	Hypertrophic
Playing	Repetitive activities = frenzied activities
Rooting	Massaging and suckling on other piglets
Standing	Nibbling on other piglets
Walking	Massaging on objects
Exploring	Deplaced eliminations
	Comfort activities as in conflict behaviour
	Timidity
	One animal laying on another (fear)
	Exploring*
	Sitting
	Lying

Information based on:

Algers (1979); Grauvogl(1972); Hafez (1962); Marx (1973); Marx et al (1977); Marx and Höpfner (1977); Metz and Osterlee (1980); Mohapeloa (1978); van Putten and Dammers (1976); van Putten (1977, 1978a, 1978c*); Signoret et al. (1975); Troxler (1980).

TABLE 5

MODIFICATIONS OF BEHAVIOUR PATTERNS OF FATTENING PIGS IN A BARREN ENVIRONMENT

Behaviour pattern Hypotrophic	Hypertrophic
Total activity	Lying
Rooting	Sitting
Walking	Standing
Playing	Deplaced eliminations
Exploring	Timidity
	Comfort activities as in conflict behaviour
	Massaging and nibbling on other pigs
	Massaging and nibbling on objects

Data based on:

Buchenauer and Grauvogl (1975); Buchenauer (1978); Etter-Kjelsaas (1980); Ewbank (1974); Haske-Cornelius (1977); van Putten (1978b, 1980).

TABLE 6

MODIFICATIONS OF BEHAVIOUR PATTERN OF SOWS IN A BARREN ENVIRONMENT

	Behaviour pattern: Hypotrophic	Behaviour pattern: Hypertrophic
Pregnant sows	Total activity	Restlessness
	Rooting	Sitting
	Walking	Standing
	Lying	Comfort activities such as conflict behaviour
	Defaecating	Unnatural lying position
	Exploring	Biting bars
		Vacuum chewing
Suckling sows	Total activity	Restlessness before farrowing
	Rooting	Lying
	Walking	Sitting
	Standing	Unnatural posture during elimination
	Nesting behaviour	
	Contact between sow and piglets	
	Defaecating	
	Exploring	

Information based on:

Algers (1979); Buchenauer (1980); Fraser (1975); Grauvogl (1958); Hafez (1962); Metz and Oosterlee (1980); Pflug (1976); van Putten (1977; 1978a,b); Schremmer and Deckert (1967); Sommer et al (1978); Studer (1975); von Zerboni di Sposetti (1977).

behaviour patterns of sows caused by the barren environment of intensive stable systems.

Pregnant sows can only carry out their nesting behaviour in a rudimentary way immediately before giving birth. Possibly the increasing restlessness may be attributed to this. Locomotor activities of pregnant and nursing sows are reduced because of the limitations in rooting, walking and exploring. Furthermore, it has been reported that these animals show changes in eliminative behaviour. Both the retention of faeces as well as defaecating and urinating in a lying position were observed. It has been reported that pregnant sows show an increased frequency of continuous sitting for up to 6 hours and this is associated with a significant drowsiness. Remaining in this position for such a long time may be attributed to frustration of the animals. Standing for long periods may be regarded as a conflict between the desire for activity and the impossibility of achieving it. Frequent comfort activities occur as displacement activities and this also implies conflict situations.

Pregnant sows adopting unnatural lying positions can be observed in tied stable systems. Because of the limited space the legs of the animal are often under the body of the neighbouring sow. Since the sows may hurt their legs when they draw their feet back they try to avoid this danger by lying on the belly with stretched fore and hind legs instead of lying on the side.

Sows which are allowed to move freely in the pen show more activity and spend more time with their piglets.

3. CONCLUSIONS

In conclusion I submit that all of the three methods described are satisfactory for assessing welfare.

I. Keeping pigs in stable systems which cause pain and fear should be avoided.

II. The frequent occurrence of behavioural disturbances are indications of frustration and may lead to suffering.

III. Modifications of behaviour show disturbance of well-being. They may also lead to suffering unless they can be offset by suppressed demand activities. However, in most housing systems such suppressed demand activities are not possible. Little is known as to what extent modification in behaviour should be tolerated. It is necessary to collect more data in this area.

REFERENCES

Achebe, Ch., 1975. Vergleichende Untersuchungen einiger Verhaltensweisen von konventionellen und frühabgesetzten Ferkeln unter Berücksichtigung des Tierschutzes, Diss. Berlin (D 83 Nr. 50).

Algers, B., 1979. Gesundheit in Flatdeck-haltung, Vortr. auf Informationstagung der Internationalen Gesellschaft für Nutztierhaltung in Basel.

Brambell, F.W.R., 1965. Report of the Technical Committee to Enquire into the Welfare of Animals kept under Intensive Husbandry Systems, Command paper No. 2836, 1-86, HMSO London.

Buchenauer, D. and Grauvogl, A., 1975. Die Aktivität von Mastschweinen in verschiedenen Haltungssystemen. In: Haltungssysteme und Verhaltensanpassung. Tagungsbericht d. Dt. Vet. Med. Gesellschaft in Freiburg KTBL Darmstadt.

Buchenauer, D., 1978. Behaviour of fattening pigs in relation to the type of floor, 1st World Congress on Ethology Appl. to Zootechnics, Madrid.

Buchenauer, D., 1980. Untersuchungen zum Verhalten von ferkelführenden Sauen in der Laufbucht und in der Abferkelbucht, Vortrag vor Fachgruppe "Verhaltensforschung" der Dt. Vet. Med. Gesellschaft, Freiburg.

Duncan, I.J.H., 1980. Can scientific research help in assessment of animal welfare? Proc. of the symposium "Behaviour In Relation to Reproduction, Management and Welfare of Farm Animals", Armidale, Austr. Sept. 1979.

Eibl-Eibesfeld, I., 1972. Grundriss der vergleichenden Verhaltensforschung, R. Piper and Co. Munich.

Etter-Kjelsaas, H., 1980. Zur Beurteilung des Offenfront-Tiefstreusystems für Mastschweine, ethologische und veterinärmedizinische Parameter, Vortrag vor Fachgruppe "Verhaltensforschung" der Dt. Vet. Med. Gesellschaft, Freiburg.

Ewbank, R., 1974. The influence of diet on general activity in fattening pigs, Proc. 3rd Intern. Pig. Vet. Soc. Congr., Lyon.

Fraser, D., 1975. The effect of straw on the behaviour of sows in tether stalls, Anim. Prod. 21, 59-68.

Grauvogl, A., 1958. Über das Verhalten des Hausschweines unter besonderer Berücksichtigung des Fortpflanzungsvermögens, Diss. Berlin (Freie Universität Journalnr. 215).

Grauvogl, A., 1972. Tierschutz aus der Sicht der modernen Verhaltensforschung, Die Kleintierpraxis 17, 181-183.

Grauvogl, A., 1978. Parameter für das Wohlbefinden bei landwirtschaftlichen Nutztieren. In: Siege, P., Bessei, W., and Grauvogl, A.: Verhaltensmerkmale bei der Züchtung von Geflügel. Hohenheimer Schriftenreihe Bd. 93, Verlag Eugen Ulmer Stuttgart.

Günster, B., 1972. Untersuchungen über das Verhalten von Ferkeln bei Einzelkäfighaltung und konventioneller Aufzucht, Diss. Hannover.

Hafez, E.S.E., 1962. The Behaviour of Domestic Animals, Baillière Tindall, London.

Haske-Cornelius, H., 1977. Untersuchungen zum Verhalten von Mastschweinen unter besonderer Berücksichtigung des Schwanz- und Ohrenbeissens in verschiedenen Stallsystemen, Diss. Vienna.

Hassenstein, B., 1974. Lern- und Spielverhalten. In: Immelmann (Ed.): Grzimeks Buch der Verhaltensforschung, Kindler Verlag.

Hinde, R.A., 1973. Das Verhalten der Tiere, Suhrkamp Verlag, Frankfurt/M.

Hughes, B.O., 1976. Behaviour as an index of welfare. In: Kilgour, 1980.

Kilgour, R., 1980. Animal welfare - the conflicting viewpoints. Proc. Behaviour in Relation to Reproduction, Management and Welfare of Farm Animals, Armidale, Austr., Sept. 1979.

Lorz, A., 1973. Tierschutzgesetz, Kommentar. In: van Putten, G. and Dammers, J., 1976.

Marx, D., 1973. Vergleichende Untersuchungen über das Verhalten von Saugferkeln und frühabgesetzten Ferkeln in Käfiggruppenhaltung (Batteriehaltung), Berl. Munich. Tierärztl. Wschr. 86, 301-306.

Marx, D., Schrenk, H.J. and Schmidborn, Ch., 1977. Spiel- und Eliminationsverhalten von Saugferkeln und frühabgesetzten Ferkeln in Käfiggruppenhaltung (Flatdecks), Dtsch. Tierärztl. Wschr. 84, 125.

Marx, D. and Höpfner, G., 1977. Verhalten frühabgesetzter Ferkel in Käfiggruppenhaltung bei verschiedenen Flatdeckausführungen. 2. Mitt. Gegenüberstellung zum Verhalten von Saugferkeln und Ferkeln in Batteriekäfighaltung, Berl. Munich Tierärztl. Wschr. 90, 358-363.

Meyer-Holzapfel, M., 1956. Über die Bereitschaft zu Spiel und Instinkthandlungen, Z. Tierpsychol. 13, 442-463.

Meijer, P., Rijnris, H.J. and van Logtestijn, J.G., 1976. Staartonsteking bij het varken, Tijdschr. Diergeneesk., 1073-1079: Müller, J. In: Teutsch, G.M., Loper, E.V., Martin, G. and Müller, J. Intensivhaltung von Nutztieren aus ethischer rechtlicher und ethologischer Sicht, Birkhauser Verlag Basle, Boston, Stuttgart (1979).

Metz, J.H.M. and Oosterlee, C.C., 1980. Immunologische und ethologische Kriterien fur die artgemässe Haltung, Vortr. vor Fachgruppe "Verhaltensforschung" der Dt. Vet. Med. Gesellsch. in Freiburg.

Mohapeloa, T.K., 1978. Tierschutz- und nutzungsbezogene Untersuchungen zur Käfigaufzucht frühabgesetzter Ferkel, Diss. Göttingen.

Pflug, R., 1976. Geburtsverhalten von Sauen und Verhaltensweisen ihrer Ferkel, KTBL-Schrift 208, Darmstadt.

van Putten, G., 1968. Een onderzoek naar staartbijten bij mestvarkens, Centrum voor landbouwpublikaties en landbouwdocumentatie Wageningen.

van Putten, G., 1976. Ferkel in einer reizarmen Umwelt. In: Probleme tierartgerechter Haltung, KTBL Darmstadt, 30-41.

van Putten, G., 1977. Die Problematik des Wohlbefindens bei Schweinen in intensiven Haltungsformen, Rapport B-127, Instituut voor Veeteeltkundig Onderzoek "Schoonoord". Zeist.

van Putten, G., 1978a. Spezielle Ethologie, Schwein. In: Nutztierethologie, (Ed. H.H. Sambraus), Verlag Paul Parey Berlin, Hamburg.

van Putten, G., 1978b. Comfort behaviour in pigs: Informative for their well-being. In: The Ethology and Ethics of Farm Animal Production,(Ed. D.W. Folsch), Birkhäuser Verlag Basel und Stuttgart.

van Putten, G., 1978c. Das sogenannte Spiel der Flatdeckferkel. Dtsch. Tierärztl. Wschr. 85, 462-464.

van Putten, G.., 1980. Objective observations on the behaviour of fattening pigs, Animal Regulation Studies 3, 105-118.

Sambraus, H.H., 1978. Analmassage und Kotfressen bei Mastschweinen, In: Aktuelle Arbeiten zur artgemässen Tierhaltung, KTBL-Schrift 240, Darmstadt.

Sambraus, H.H., Schunke, B. and Kräusslich, H., 1980. Verhaltensanomalien von Zuchtsauen in Kastenständen, Vortrag auf DGfZ/GfT-Tagung in Berlin.

Schremmer, H. and Deckert, R., 1967. Untersuchungen über das Verhalten und die Leistung säugender Sauen in Anbindehaltung, Arch. Tierzucht 10, 399-409.

Schrenk, H.-J., 1978. Einfluss von Licht auf den Tagesrhythmus der Aktivität von Ferkeln. Personal communication.

Signoret, J.P., Baldwin, B.A., Fraser, D. and Hafez, E.S.E., 1975. The Behaviour of Swine. In: The Behaviour of Domestic Animals (Ed. E.S.E. Hafez), Baillière Tindall, London.

Smidt, D., Andreae, U. and Unshelm, J., 1980. Ist "Wohlbefinden" messbar? - Anmerkungen zu einem Tierschutzproblem - Tierzüchter, 338-340.

Sommer, B. and Kräusslich, H., 1978. Zuchtsauen in Kastenständen, Vortr, auf Weltkongress für Tierschutz, 26, 29.9. in Berlin.

Studer, H., 1975. Das Verhalten von Galtsauen in verschiedenen Aufstallungssystemen, Diss. Med. Vet. Bern.

Troxler, J., 1980. Ethologische und veterinärmedizinische Parameter bei der Beurteilung zweier Haltungssysteme fur Absatzferkel, Vortr. vor Fachgruppe "Verhaltensforschung" der Dt. Vet. Med. Gesellsch. in Freiburg.

Tschanz, B., 1978. Reaktionsnorm und Adaptation. Das Tier im Experiment, 33-49.

Weiss, M., 1975. Lernspezifische Einflüsse im Erkundungs- und Spielverhalten von Iltisfrettchen (*Mustella putorius x M. furo*), Diss. Marburg.

Wood-Gush, D.G.M., Duncan, I.J.H. and Fraser, D., 1975. Social stress and welfare problems in agricultural animals. In: The Behaviour of Domestic Animals (Ed. E.S.E. Hafez), Baillière Tindall, London.

von Zerboni di Sposetti, H.N., 1977. Untersuchungen zum Verhalten von Zuchtsauen in unterschiedlichen Aufstallungsformen, Diss. Vienna.

DISCUSSION

R. Moss *(UK)*

In Table 1, what exactly are you taking as the criteria of the infliction of pain, and what in fact is being used to promote fear, or are these generalisations of what you expect is fear and what you expect is painful?

Doris Buchenauer *(FRG)*

I expect pain when the pig is in the situation of feeling hurt, and I expect fear as an expression of emotional pain.

R. Moss

Your criteria, then, for what is fearful to a pig is what you would expect to be fearful from a human point of view?

Doris Buchenauer

No, we tried to get the information from the reactions of the pig, not from our own ideas of fear. Our data are as objective as possible.

I. Ekesbo *(Sweden)*

Can you clarify the results in Table 2, 'Anomalies in the behaviour of pigs'? Are you giving average figures for the incidence of the various behavioural activities in the four groups?

Doris Buchenauer

In this Table, I want to show that these disturbances appear in all development stages for the animal. I have used data from the literature that prove these observations statistically. I also believe that hyperactivity is found in suckling piglets reared in poor environments, but I could not find any data to support this observation.

R. Dantzer *(France)*

First, you have talked about hypotrophic versus hypertrophic. Does this mean that in some cases there is a decrease in behaviour, and in other cases there is an increase in behaviour?

Doris Buchenauer

Yes, that is correct.

R. Dantzer

Secondly, I want to return to the point made at the beginning of this discussion. What are your criteria of reference by which you describe firstly the reaction to pain in pigs, and secondly the reaction to fear in pigs?

Doris Buchenauer

This item is not based on my own observations but upon data from Grauvogl. But, I am sure that I can make my own observations to sustain the points made on the expression of pain and fear in pigs. Reactions can be seen to annoyance - changes in behaviour or position.

R. Dantzer

Yes, but do you have a situation where you can manipulate the level of fear of the animal so as to be sure that your criteria are correct?

Doris Buchenauer

No, this is really an hypothesis.

B.W. Moss *(Northern Ireland)*

I would like to suggest that tail-biting is started by a particular pig, one that could be called a 'criminal' pig, and that it is not abnormal behaviour brought on by environment. Environment merely exaggerates the behaviour. I have observed that this vice occurs in pens that were uncrowded and that particular pigs were responsible, which contrasts the normal view that it is a vice that occurs in intensive situations. Have you any observations to support or reject this?

Doris Buchenauer

I would not agree completely with you because most observations in the literature have found that cannibalism occurs more frequently in a poor environment, although this does not mean vices such as tail-biting cannot be found in other environments. Lindquist in Sweden made observations on 155 farms and found that the vast majority of cases where tail-biting occurred were in poor environment situations.

D.L. Ingram *(UK)*

I would like to make a general observation on tail-biting and nose rubbing. If you give a pig the drug apomorphine you can turn this behaviour on with 100% reliability, but the extent to which they rub or bite depends a little on the individual pig. So perhaps what we are dealing with is an abnormality in the dopamine metabolism in the brain, and it would seem to me that some pigs are very much more susceptible to it than others.

Doris Buchenauer

This abnormality in the brain might be caused more often in this environment?

D.L. Ingram

Indeed, the same applies to the early-weaned pig which is already showing signs of changes in the brain metabolism that could, presumably, predispose it to this sort of behaviour.

G. van Putten *(Netherlands)*

I do not believe tail-biting to be caused by one 'criminal' animal. It is more a case of redirected explorative behaviour.

If one examines how it starts and develops then it is surely not aggressiveness. The animals begin nibbling each other harmlessly and quietly, and it would seem to have much in common with explorative behaviour. There is a theory to explain why this would happen more often in a barren environment than in a rich environment. A barren environment makes it necessary for the animal to explore more because it has nothing else to do. I cannot agree with Dr. Moss that one 'criminal' animal is to blame, although he is right in one way when he says that it mostly begins with one animal - the smallest animal. But the smallest animal might be more susceptible to a barren environment.

B.W. Moss

In connection with your comment about the relation of tail-biting to exploratory behaviour, what would the position be in relation to a lower space allocation per pig, where the space for exploration is decreased? What influence would that have on tail-biting?

G. van Putten

A higher density of animals per square metre is not in itself a reason for cannibalism, but it does cause a bad atmosphere in the stables. We know that bad ventilation increases explorative behaviour and thereby might increase such activities as tail-biting. However, I did not state that a higher density of animals, in itself, causes tail-biting.

I. Ekesbo

I agree with Dr. van Putten, and yet I would like once more to raise the point about the 'criminal' pig. High density, barren environments can act to over-stimulate the animal. If animals are kept on slatted floors with no straw then high noise levels could also over-stimulate the animal. There is both under- and over-stimulation in a barren environment which can help cause the symptoms of tail-biting. However, some individuals show these symptoms in quite rich environments where there is little under- or over-stimulation, and it does not spread to other animals. It occurred before we ever had intensive animal houses, and the term 'criminal' is often used by the farmers' themselves. I have never seen this behaviour pattern described in the literature, but I would guess that it is a symptom that occurs in maybe 1 out of 1 000, or even 5 000, pigs.

LOSSES DUE TO INHERITANCE AND HOUSING AS CRITERIA FOR THE WELFARE OF PIGS

J. Unshelm
Institute of Animal Husbandry and Animal Behaviour Mariensee,
Agricultural Research Centre Braunschweig-Völkenrode,
Federal Republic of Germany.

1. INTRODUCTION

There are several possible ways of assessing animal welfare problems arising from breeding and management. The following aspects should be observed:

1. the behaviour of animals,
2. physio-biochemical stress parameters,
3. clinical dignoses,
4. patho-anatomical changes,
5. percentage losses.

In making these observations, there must be a differentiation between the genetically influenced performance level - which is equal to the susceptibility of the population - and that which is influenced by management. The parallel determination of all these parameters, and in particular ethological and physio-biochemical ones, is part of the concept on which our Institute is based. Extensive investigations of this kind are an important aid in working out rules, as we have to supply information to assist the creation of directives necessary to ensure better application of the animal welfare act.

In future, animal welfare laws and related rulings will have to show how far behavioural anomalies, stress reaction, clinical diagnoses, patho-anatomical changes and death losses can be accepted. It must, however, be realised that a death loss rate of zero is practically impossible. This is a most important point in the evaluation of the parameters which have been mentioned, and I will refer to it again at the end.

Having given some information about ethological and physiological criteria as measures for species-specific management of pigs, I would like to elaborate on the subject of losses as a means of evaluation.

2. EVALUATION OF DIFFERENT KINDS OF LOSSES

As has already been mentioned, we can differentiate between

1. predominantly genetic causes (i.e. breed differences)
2. exogenous factors (e.g. management systems, herd size, etc.)

In addition, we may subdivide these two groups into the categories of rearing, i.e. piglets, fattening and transportation.

First, I will deal with losses during rearing, i.e. losses of piglets.

2.1. Losses of piglets

TABLE 1

PIGLET LOSS IN RELATION TO LITTER NUMBER*

Litter number	Piglets born	Piglets weaned	Losses (%)
1	8.9	8.1	8.5
2	9.8	8.9	8.5
3	10.3	9.3	9.2
4	10.4	9.4	9.9
5	10.4	9.4	10.6
6	10.4	9.3	11.0
7	10.3	9.1	11.4
8 and over	10.0	8.8	11.7

* n = 157 112 litters

Schweineproduktion 1978

Table 1 shows the relationship between the number of litters a sow bears and the amount of piglet loss. This is a trait which cannot be influenced. From the 157 112 litters studied, we can see that the more litters a sow bears, the more piglets are lost per litter, whilst the number of piglets born decreases after the sixth litter. There is, however, a certain amount of selection taking place because nobody will keep a sow after her seventh or eighth litter if the number of piglets born or reared is insufficient.

2.2. Rearing systems

TABLE 2

PIGLET LOSS IN RELATION TO GROSS MARGIN* AND REARING SYSTEM

Gross margin	Rearing system	Losses % 1975	1976
Low	Conventional	15.8	15.3
High		11.0	11.6
Low	Flatdeck	17.2	14.3
High		13.2	13.0

* Margin/sow/year, comparison between 25% of the best and 25% of the poorest results (out of a total of approximately 1 800 farms).
Schweineproduktion 1976 and 1977.

Table 2 compares the rearing system, divided into conventional and flatdeck rearing, and the gross margin, which is the income/sow/year minus the costs/sow/year. The comparison was made between the 25% of farms registered which had the best results and the 25% which had the poorest results. The wide variation was due principally to different management practices. Taking this as a starting point, the Table shows that farms with good management lose less piglets than farms with poor management, irrespective of which rearing system is used. There is no doubt that this is a positive aspect because it indicates a positive relationship between economic interests and the welfare requirements of the animals.

There is an obvious interaction between the rearing system and the gross margin. Well managed farms suffer greater losses with flatdecks than with conventional rearing, whereas in farms with a low gross margin large losses can be observed with both systems, varying from one year to another. Decreasing losses with the flatdeck system on farms with a low gross margin could be interpreted as an indication that there is a learning process involved in using this system. This tendency can also be observed in farms with good management.

TABLE 3

PIGLET LOSS IN RELATION TO GROSS MARGIN* AND HERD SIZE (SOWS/FARM)

Gross margin	Sows/farm	Losses %
Low High	< 20	15.5 11.1
Low High	21 - 40	15.1 11.1
Low High	41 - 60	15.0 11.2
Low High	> 60	19.5 12.0

*Margin/sow/year; comparison between 25% of the best and 25% of the poorest herds (from a total of 1500 farms)

The relationship between gross margin and piglet loss which has already been mentioned can be clearly seen in Table 3, as can other forms of interaction. Farm size has no effect on piglet loss in farms with high gross margins, where the variation is from 11.1 to 12%. In farms with low gross margins, however, piglet loss increases to 19.5% if the herd size exceeds 60 sows. There is a clear limit here which is of particular interest, in that there is no difference in terms of piglet loss between farms with good or poor management as long as herd size remains below 60 sows; only when herd size exceeds 60 sows in farms with poor management is there a significant increase in piglet loss. Thus there would seem to be an area of overproportional increase in stress. As yet, it is not possible to say whether this difference is due to the fact that smaller farms are managed mainly by families with a personal interest in the animals, whereas large farms have to hire labour, or whether there are yet other factors which are important.

Further influences can only be summarised. No significant influence on piglet loss as a result of the management of pregnant sows (e.g. pasture during summer, keeping sows outside, housing in groups or individually) can be observed. Giving large amounts of bedding to pregnant sows does, however, result in significant decreases in piglet loss.

2.3. Early weaning

A decrease in piglet loss can also be observed in farms which have an 'eros centre'. Early weaning - before the 35th day of life - increases the gross margin, but at the same time there is an unimportant increase in piglet loss. Piglet loss did not decrease beyond the weaning age of 42 days.

2.4. Genetics

Indications of genetic differences can be seen from the differences between breeds. Losses are relatively low in German Large White, much higher in German Landrace and quite high in Belgian Landrace and Piétrain. The same tendency can be seen within the German Landrace, as there are more losses among piglets with a greater market weight. The result of using hybrid pigs is a higher gross margin/sow, but this is only true if the higher cost of buying the young sows is not included. The influence on piglet loss was not significant. In this respect, however, there are a number of conflicting results.

Having discussed piglets and the rearing period, we will now continue with a discussion of the fattening period. One very striking exogenous factor is the well known fact that fattening pigs bought from a dealer register greater losses than those bought directly from a well known and registered farm. With regard to housing systems, several investigations indicate that slatted floors lead to an increase of 1 - 2% in losses. There is however more information on losses available from progeny testing stations, where the reasons for loss are registered. Loss variation over the last 20 years and the influence of breed and other genetic factors are also known.

2.5. Testing stations

The total number of losses recorded in German testing stations is shown in Table 4.

It can be seen from the Table that total loss increased from 2.85% to more than 6% between 1960 and 1962, and that a high loss rate has existed since 1970. It is remarkable that a considerable percentage increase is due to disturbances of the cardio-vascular system and to leg weakness. Respiratory disease shows a decrease whilst diseases of the gastro-intestinal tract remain constant at a low level. Developmental disorders and weight deficiency also cause loss, and are determined by daily gain. A fattening pig is considered to have a developmental disorder if its daily gain remains below 400 g for a period of 3 weeks except in the Piétrain breed where gain is limited to 300 g. This type of loss was found in one third of all cases between 1970 and 1972. It therefore seems that a more exact analysis of this category would be worthwhile.

2.6. Breed differences

A breed comparison (Table 5) shows the following facts: as far as total loss is concerned, the German Large White is at one extreme of the scale, with a level of loss of 4.18%, whilst the Piétrain is at the other extreme, with a level of 8.32%, nearly twice that of the German Large White. Belgian Landrace pigs have a loss level of 6.58% and German Landrace pigs have a loss level of 6.02%, both of which figures still seem rather high. In the German Large White, losses due to cardio-vascular disturbance and leg weakness are remarkably low, whilst in the Piétrain pig a large percentage of losses are due to developmental disorders. I should like to mention here that causes of

TABLE 4

CAUSES OF LOSS IN TESTING STATIONS (%)

Causes of loss	1960 - 1962	1963 - 1965	1966 - 1968	1970 - 1972	1974 - 1976	1977 - 1979
Cardio-vascular disorders	0.60	1.10	1.12	1.26	1.38	1.63
Leg weakness	0.36	0.59	1.01	1.82	1.60	1.65
Diseases of the respiratory system	0.60	0.30	0.28	0.28	0.27	0.17
Diseases of the gastro-intestinal tract	0.15	0.19	0.26	0.18	0.14	0.16
Developmental disorders and short weight	0.69	0.68	0.82	2.15	1.62	1.57
Accidents and other causes	0.46	0.58	0.44	1.06	0.69	0.95
Total	2.85	3.36	3.93	6.73	5.62	6.13

Flock, 1969; Pflelderer,1973, 1976 (completed with new data).

loss could be analysed much more exactly and that the present praxis may lead to misinterpretations.

TABLE 5

CAUSES OF LOSS IN DIFFERENT BREEDS IN TESTING STATIONS IN 1978

Causes of loss	Breed			
	German Landrace	German Landrace B	Piétrain	German Large White
Cardio-vascular system %	1.60	1.86	1.61	0
Leg weakness %	1.59	1.66	1.10	0.38
Development disorders* %	1.58	1.98	3.44	1.98
Total	6.02	6.58	8.32	4.18

* If daily gains are less than 400 g (Piétrain 300 g) over a period of more than 3 weeks.

2.7. Halothane test

The halothane test has already been introduced into some countries and is being used on an experimental basis in others with the aim of detecting individuals and populations which are particularly susceptible to stress, and of searching for a solution. It is therefore of interest to investigate the possibility of reducing loss by using such methods. As was to be expected, and can be seen in Table 6, losses in halothane-positive animals were 7.6%, markedly higher than those in halothane-negative pigs, where losses were only 4.7%. In addition, it is interesting to note that none of the losses in halothane-negative pigs were due to cardio-vascular disorders. The percentage of skeletal defects was not influenced by halothane reaction.

I presume that more detailed results are available from countries such as the Netherlands, for instance, where the halothane test is used much more extensively than in our country.

TABLE 6

COMPARISON OF LOSSES AND HALOTHANE REACTION IN GERMAN LANDRACE PIGS

	Halothane reaction +	Halothane reaction -
Number of animals	625	235
Losses (%)	7.6	4.7
Due to:		
Deficiency of the cardio-vascular system	3.0	0
Deficiency of the skeleton	3.1	3.8
Other causes	1.5	0.9

Petri et al., 1979.

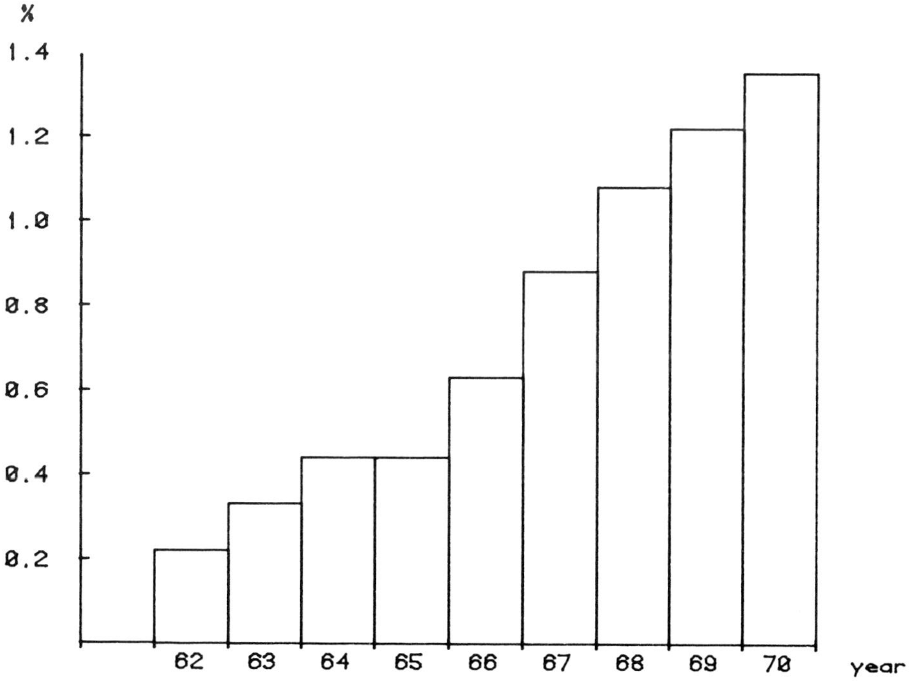

Fig. 1. Percentage death losses during transportation in Munich out of a total of 250 - 300 000 pigs/year. (Löhr, 1970).

2.8. Transportation

Figure 1, like the Tables, shows only a trend. In this Figure we see the percentage of pigs which died during transportation to the slaughterhouse in Munich between 1962 and 1970. Unfortunately, more recent data are not available. It can, however, be assumed that at least 500 000 pigs/year in the Federal Republic of Germany die on their way to slaughter. The reasons for these losses are not fully known, but some information can be gained from more accurate analyses which are made occasionally, the results of which can be seen in the meat inspection data shown in Table 7.

TABLE 7

RESULTS OF MEAT INSPECTION (%) IN PIGS FROM PROGENY STATIONS

Diagnosis	German Landrace	German Landrace B	Pietrain	German Large White
Disorders of the lungs %	7.2	6.1	3.1	0
Pericarditis %	0.2	0.2	0.2	0
Liver disorders %	3.6	3.2	2.8	0
Rachitis %	0.6	0.8	0.3	0.4
Leg weakness %	8.3	8.3	5.5	1.1
Claw abnormalities + joint disorders %	2.4	2.5	0.8	1.1

n = total ca. 17 000
Schweineproduktion, 1976

2.9 Meat Inspection

Table 7 gives the meat inspection results for different breeds. As was mentioned earlier, the German Large White showed the best results. A great number of lung disorders as well as liver defects were found in both German and Belgian Landrace breeds. These same two breeds also showed a high percentage of leg weakness, more than 10%.

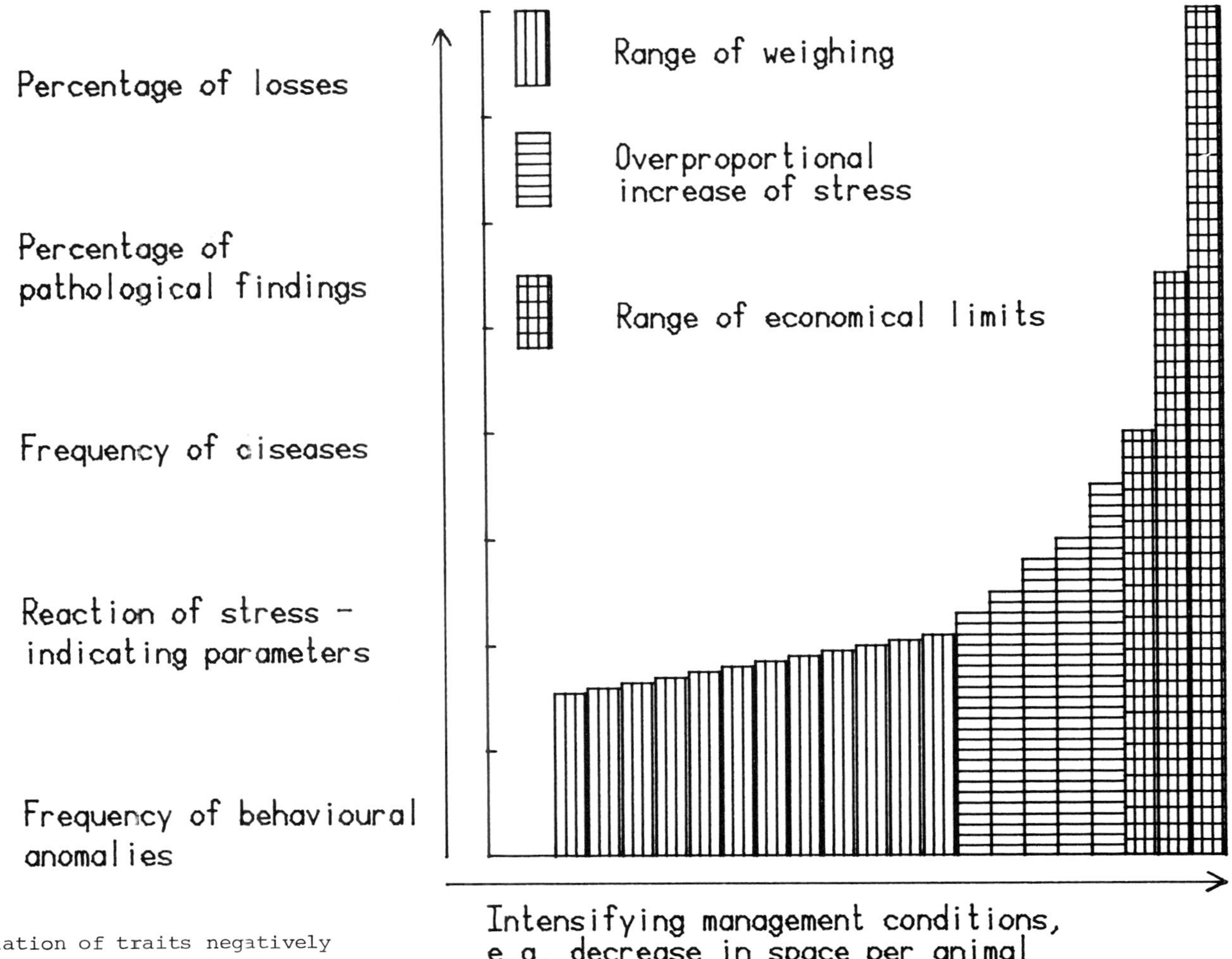

Fig. 2. Accumulation of traits negatively related to stress resistance.

2.10. Model approach

I should now like to make some observations on the consequences resulting from the figures and their interpretation. Figure 2 is a model. It is based on the hypothesis that intensifying management conditions (for example, by decreasing the space/animal) and likewise strengthening traits which are negatively related to stress resistance beyond a certain limit will result in behavioural abnormalities, a pronounced susceptibility to stress, higher morbidity and eventually greater losses. I presume that the increases will not be linear. At first changes will take place at a slow rate. This phase is shown in lines and is assumed to be the weight range according to personal judgement. Authorities, either scientists or politicians, enacting laws in accordance with scientific results have set a weight range limit to decide whether - for example - a loss of some 2% during the fattening period is acceptable whilst a loss of 3 - 4% surpasses the borderline as laid down by the law. Similarly, a decision will have to be made regarding the ceiling of plasma cortisol concentrations which result from certain housing conditions. These authorities will also have to judge whether behavioural abnormalities of individual animals are acceptable, or if one single abnormality shown by one animal should be a deciding factor.

A more intensified production scheme, or an increase in performance, which should accompany increased susceptibility to stress, leads into the cross striped area on the model, where a disproportional increase of stress is obvious. It is our duty to determine the starting point of disproportional stress and make a picture of the consequences, this being necessary before laws can be established. This starting point has to be evaluated for different populations, different breeds, which vary in susceptibility, for different schemes of management, for various situations and interactions which derive from these factors. There are, for instance, conditions under which a simple, low performance breed may not suffer at all, but under which more susceptible populations may suffer large losses. Unfortunately, a reference system, where a rather primitive breed is kept under simple conditions, is not available. The behaviour, stress reaction or frequency of disease and loss could otherwise have been used as a basis for judgement. However, care must be taken even here. It must be remembered that about 50 years ago, even in quite resistant breeds under simple production conditions, piglet losses of the order of 30% were common.

The third section of the model, marked by the pattern of checks, describes the range of economic limits, where measures would have to be taken for economic defence. Under these conditions it can be assumed that certain production conditions and performance levels would result in extremely large losses, which in turn would negatively influence the economy of pig production. At this point conditions would have to be changed. Obviously, we no longer want to think about this area in connection with animal welfare. We should aim rather to develop tools which will enable us to determine the beginning of the disproportional increase in stress and to think about whether we are already able, now, to make recommendations for limiting the scope of personal judgement. There is no doubt that in this respect the interests of animal welfare and economics may lie in the same direction.

3. CONCLUSIONS AND RECOMMENDATIONS

I. Pathological studies, as well as physiological and ethological investigations, should be used to find the limits to certain production schemes and situations.

II. If the limits of evaluating parameters are exceeded under certain production conditions, then these conditions must be changed or replaced.

III. Breeds of pigs, pig populations and hybrid groups in which losses during rearing, fattening and transport surpass certain levels, should be culled and replaced by more resistant animals. This is in the interests of both animal welfare and economy.

REFERENCES

Andreae, U., Unshelm, J. and Smidt, D., 1978. Zeitgleiche ethologische und verhaltensphysiologische Studien an Mastbullen Tierzüchter 30, 332-334.

Flock, D., 1969. Untersuchungen über Ausfallursachen und Gesundheitsbefunde bei der Stationsprüfung von Schweinen Dtsch. tierärztl. Wschr. 76, 429-433.

Löhr, J., 1970. Mortality in pigs due to transport. Proc. Symp. 'Stress in the pig', Beerse/Belgien.

Petri, W., Oster, W., Haberkorn, H., Gindele, H.R. and Köppen, W.D., 1979. Untersuchungen zur Streßanfälligkeit beim Schwein Tierzüchter 31, 496-498.

Pfleiderer, U.-E., 1973. Ausfallsursachen bei Mastprüfungsschweinen der Deutschen Landrasse. Schweinezucht und Schweinemast 21, 220 and 222.

Pfleiderer, U.-E., 1978. Konstitutionsschwäche und Leistung Schweinezucht und Schweinemast 26, 329-331.

Scheven, B. and Ingwersen, J., 1975. Schweineproduktion in der Bundesrepublik Deutschland. Kommentator-Verlag GmbH, D 6000 Frankfurt/ Main 97.

Scheven, B. and Ingwersen, J., 1976. Schweineproduktion in der Bundesrepublik Deutschland. Kommentator-Verlag GmbH, D 6000 Frankfurt/ Main 97.

Scheven, B. and Ingwersen, J., 1977. Schweineproduktion in der Bundesrepublik Deutschland. Kommentator-Verlag GmbH, D 6000 Frankfurt/ Main 97.

Scheven, B. and Ingwersen, J., 1978. Schweineproduktion in der Bundesrepublik Deutschland. Kommentator-Verlag GmbH, D 6000 Frankfurt/ Main 97.

Smidt, D., Andreae, U. and Unshelm, J., 1980. Ist 'Wohlbefinden' meßbar? - Anmerkungen zu einem Tierschutzproblem - Tierzüchter 32, 338-340.

Unshelm, J., 1971. Konstitutionskriterien bei Schweinen verschiedener Rassen. Habil.-Schrift Göttingen 1970, Schriftenreihe des Max-Planck-Instituts für Tierzucht und Tierernährung, Sonderband.

Unshelm, J., 1976. Konstitutionsprobleme beim Schwein. Landbauforschung Völkenrode 26, 97-101.

Unshelm, J., 1980a. Verhaltensphysiologische Indikatoren für tierschutzgerechte Haltung von Mastbullen und Mastkälbern Landbauforschung Völkenrode, Sonderheft 53, 74-85.

Unshelm, J., 1980b. Verhaltensphysiologische Aspekte der Bullenmast Hohenheimer Arbeiten Heft 107, Verlag Eugen Ulmer, D 7000 Stuttgart 94-110.

Unshelm, J., Smidt, D., Andreae, U., Ellendorff, F. and Elsaesser, F., 1978. Haltungssysteme und soziale Rangordnung als Einflußfaktoren biochemischer Parameter in 'Aktuelle Arbeiten zur artgemäßen Tierhaltung 1977', KTBL-Schrift 233, Landwirtschaftsverlag GmbH, D 4400 Münster-Hiltrup, 179-185.

- Berichte aus Verden 1979. Rechenzentrum zur Förderung der Landwirtschaft in Niedersachsen, Heideweg 1, D 2810 Verden.

DISCUSSION

R. Moss *(UK)*

What criteria were used for 'good' and 'poor' management, and what are the significant factors within those two criteria which relate to the figures that were produced?

J. Unshelm *(FRG)*

We registered the gross margins of many farms. The gross margin is the income minus the cost/sow/year. Then we compared 25% of the best farms against 25% of the poorest farms. So we took the two extremes.

R. Moss

So you are using gross margins as a criterion of management of livestock; is that not in fact a criterion of the use of capital investment and not necessarily a reflection of the management of the stock itself.

J. Unshelm

No, the difference is in the economy of the individual farms.

R. Moss

Surely, that does not necessarily reflect good management of the stock in terms of the way the stock are handled, from a welfare point of view, it is more an economic viewpoint.

J. Unshelm

Not all the economic consequences are related to the gross margin, but certainly some of them.

A. Hoogerbrugge *(Netherlands)*

I have a question about the average losses (of about 10 and 11%) shown in Table 1. This is very low in comparison with figures from Sweden, Denmark, the UK and Holland which reveal losses of about 15 or 16%. Did you begin to count these losses after one or two days or did you start directly after birth? Why are the figures for Germany so different?

J. Unshelm

The figures are from farms which specialise in piglet production, not from all the farms in Germany.

A. Hoogerbrugge

However, it is still an enormous number of litters; it is very difficult to collect reliable figures.

J. Hanrahan *(Ireland)*

In your paper you mention that large amounts of bedding for pregnant sows results in a significant decrease in the loss of piglets. Is this referring to pregnant or suckling sows?

J. Unshelm

This refers to pregnant sows.

THE DEVELOPMENT OF A BLOOD PROFILE FOR STRESS ASSESSMENT

B.W. Moss

Department of Agriculture for Northern Ireland and
Faculty of Agriculture and Food Science,
Queen's University of Belfast, Northern Ireland.

1. INTRODUCTION

It has been suggested that the subject of stress be approached with a 'modicum of anthropomorphism' (Napier, 1973), however, the attribution of human feelings to animals can lead to some confusion, particularly when interpreted from behaviour studies (Bareham, 1973; Fraser, 1973). The subjective confusion which surrounds most anthropomorphic arguments may be best overcome by an objective quantitative assessment of stress from physiological measurements. Such a system based on blood profiles, however, must not be completely devoid of anthropomorphism. 'Stress' is a condition experienced by humans and, on the grounds of comparative physiology, there is ample evidence to indicate that stress responses can be attributed to animals.

2. STRESS CONCEPT

The concept of stress first developed by Selye in 1951, particularly the role played by pituitary/adrenal activation, has influenced most of the subsequent work in this area. However, one of Selye's more important concepts, that of describing diverse noxious stimuli as stressors, and then considering their responses, appears to have been overlooked in later literature. Ewbank in 1973 incorporated this idea of stressor and response in a useful concept of stress which included the conditions of overstress and understress. The importance of separating the stressor and its response becomes evident when the following, not uncommon, statement is considered: 'pigs crowded in intensive systems are more stressed'. The first obvious ambiguity is whether crowded means lower space allocation or greater group size, but what is more important is the term stressed. Does this statement mean that the pigs are subject to a higher stressor level due to close physical contact, possibly coupled with restriction of movement, or the pigs' response is such that they could be described as suffering stress? It is important to distinguish between stressor and response in order to assess and interpret stress. Blood profiles may in fact help to make this distinction.

3. BLOOD PROFILES

The first problems encountered in using blood profiles will be practical ones such as:-

1) how to restrain the pig

2) selection of the site for blood sampling; and

3) the timing to sampling in relation to diurnal variation or husbandry practice.

It is inevitable that some stress will occur during the actual sampling and this must be assessed by behavioural measurements or by appropriate parameters in the blood profile. Ideally we must be able to determine whether the levels of such parameters are due to the inherent responsivity of the animal or the amount of stressor imposed during sampling. In this respect, it is relevant to consider the types of blood parameter responses that may be observed. Basically two types of responses can be identified:

1) those in which the blood level is controlled by a feed-back system e.g. hormonal response and

2) those in which blood levels are controlled by metabolism or excretion e.g. serum enzyme activities and free fatty acids.

At any one time after a given stressor, the circulating level of any parameter in the blood will depend on its rate of release and its rate of removal.

Compartmental models can be constructed to predict the responses of blood parameters to changes in release or removal rate constants. Two types of models are proposed, one for the feedback or adaptive response and the other for a response which has no feedback system (non-adaptive response). In this context the use of adaptive refers to changes in stimulus-response sensitivity at the appropriate target organ. In the case of the hormonal system, there may be several target organs e.g. hypothalamus, pituitary and adrenal glands for corticosteroid responses. Figure 1 shows the compartmental models proposed for the two systems.

The major difference between the two models is the existence of a feedback system in the adaptive response. It is difficult to produce an accurate mathematical model incorporating a feedback mechanism as there may be a continuous feedback system operating or one which only operates when a threshold value is reached. For modelling purposes, negative feedback can be considered as decreasing the rate of secretion or as increasing the rate of removal.

In this paper two approaches have been used to simplify the problems of feedback. The first approach (Figure 2) is to consider that the rate constants for stressor stimulated secretion (k_a), feedback inhibition (k_f), and removal (k_c) are operative from the initiation of the stressor and remain unchanged throughout the period of stressor application. In the second approach, illustrated in Figure 3, the secretion due to stressor stimulation is discontinued at a certain time after the initiation of the stressor (time at q max in Figure 3).

In physiological terms this cessation of secretion may be due to:

1) secretion (k_a) continuing only whilst the stressor is applied i.e. time to q max measures the duration of the stressor, or

2) feedback inhibition switching off secretion i.e. q max may not be closely related to intensity of the stressor.

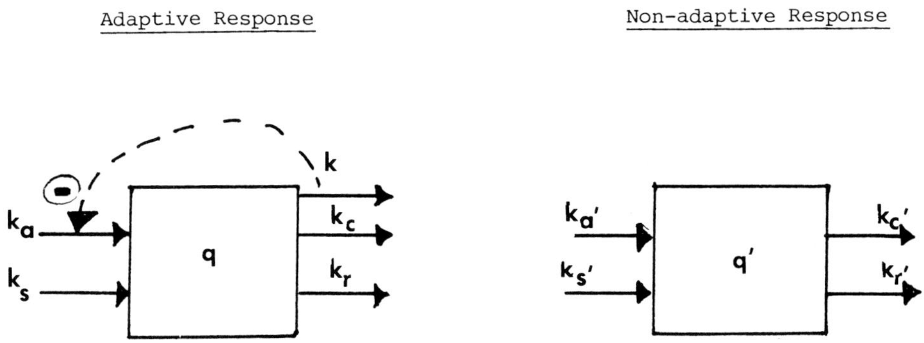

q -	quantity of parameter	- q^1
k_s -	rate of secretion in resting state	- $k_s{}^1$
k_r -	rate of removal due to metabolism and/or excretion	- $k_r{}^1$
k_a -	rate of secretion due to stressor stimulation	- $k_a{}^1$
k_c -	rate of removal due to mass action effect	- $k_c{}^1$
k_f -	represents feedback inhibition (→Θ) on k_a	

NOTES

1 Feedback inhibition (k_f) actually reduces k_a but for modelling purposes can be considered as an increase in the rate of removal.

2 For many types of non-adaptive response $k_a{}^1$ is much greater than $k_s{}^1$, in fact $k_s{}^1$ may be almost undetectable in a true resting state.

Fig. 1. Proposed kinetic models for feedback (adaptive) and non-feedback (non-adaptive) systems.

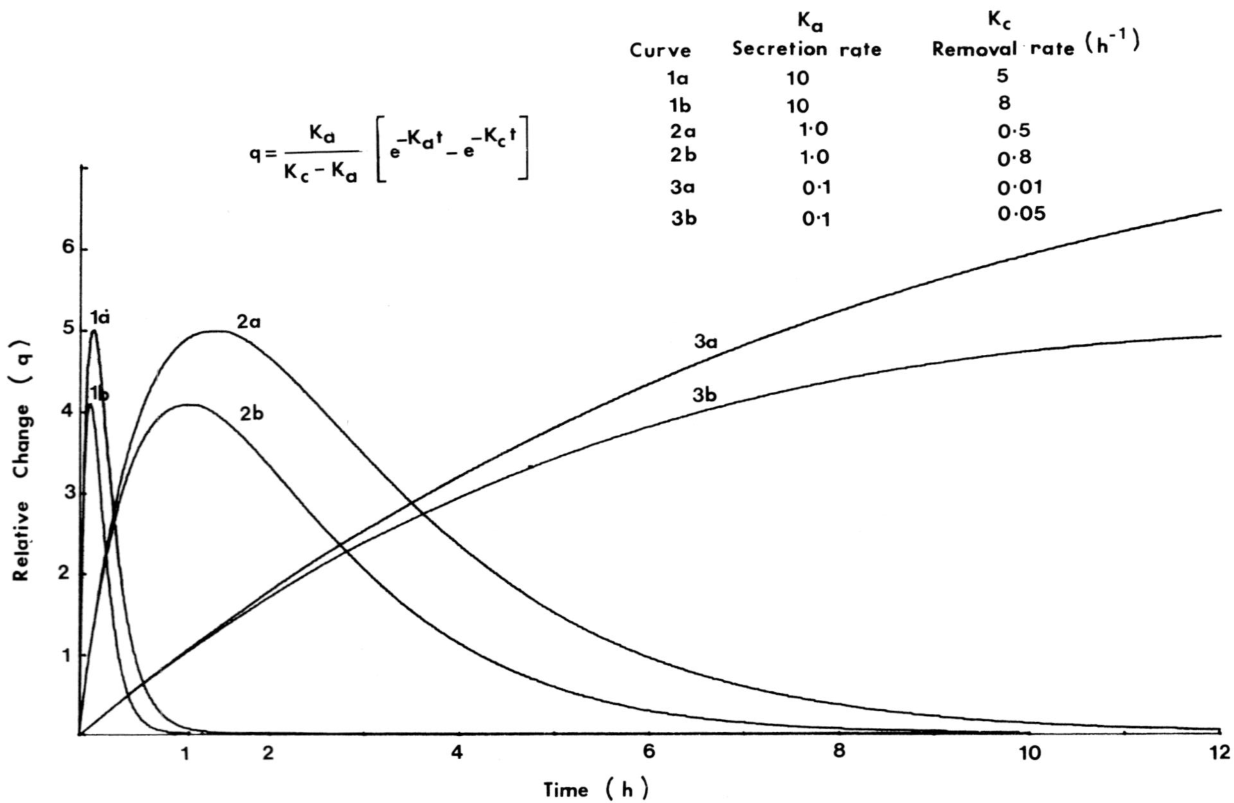

Fig. 2. Predicted responses to stressors.

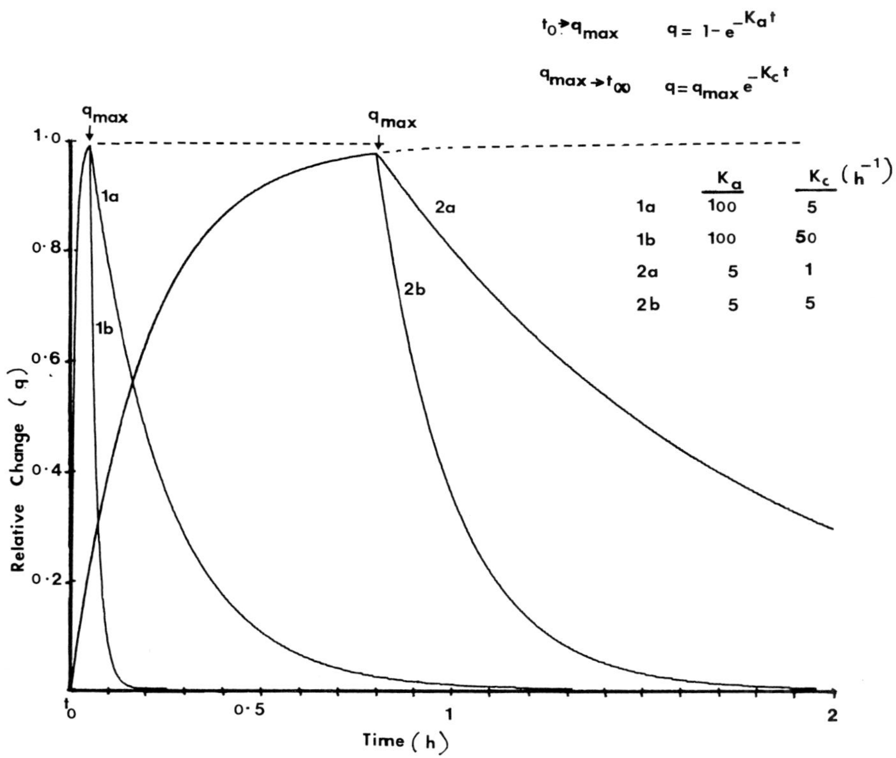

Fig. 3. Effect of duration of stressor/response

The other difference to note between the two systems is that the normal rate of secretion (k_s') in the non feedback system is generally low, and in a true resting state the parameter may be almost undetectable in the blood. Similarly the removal rate constants (k_r') and to some extent (k_c') will also be relatively small and rates of removal may control blood levels more than rates of secretion.

The compartmental models while essentially simple simulate the many physiological interactions which occur. For example, curve 1 can be considered typical of adrenal catecholamine responses; curve 2, adrenal corticosteroid responses; and curve 3, serum CPK (creatine phosphokinase) responses, to a given stressor.

4. PRACTICAL IMPLICATIONS

The practical implications of the various response curves are as follows:-

1) the faster the relative rate of change of the parameter in the blood, the greater the effect of variations in sampling time on that parameter;

2) a corollary to (1), at a given time the variation in the levels of a parameter between animals reflects differences in sensitivity i.e. relative rate of change;

3) the capacity of an animal to respond to a given stressor can be best identified by measuring the parameter at the time when maximum response has been reached;

4) measurement of parameters which exhibit type 3 curves should not be significantly affected by the sampling stressor;

5) parameters with type 3 curves can be used to give an indication of the previous stressor history.

If the kinetics of the model proposed in Figure 2 can be obtained by physiological measurements (Sebranek et al., 1973; Lister et al., 1972; Marple and Cassens, 1973) a precise interpretation of the stressor response, and consequently the stress status of the animal can be deduced.

For practical purposes, a blood profile to measure stress will be composed of the following:-

1) an assessment of sympathetic activity e.g. catecholamine levels or dopamine-β-hydroxylase activity;

2) an adaptive response e.g. corticosteroids or thyroxine; and

3) a non adaptive response e.g. serum CPK or LDH (lactate dehydrogenase) activities.

In interpreting the responses to the various stressors, catecholamine levels would be used to indicate the stressor involved in sampling; corticosteroid levels would show the amount of adaptation to long term stressors and also, to a certain extent, the stressor level during sampling; and serum enzyme levels would indicate the amount of stressor during the previous 12 or 24 h.

5. ADAPTATION

One of the major problems in assessing stress in intensive husbandry systems is the ability of the pigs to adapt to long term stressors. For example, in an experiment where pigs were kept at various stocking rates, from 28 kg liveweight to slaughter at 80 kg, the plasma cortisol levels obtained from the ear-vein of pigs at 70 - 75 kg liveweight did not show marked differences between groups (Table 1). When the pigs were slaughtered at 80 kg, however, there were marked differences between groups, which would indicate differing responses to the stressors of pre-slaughter handling. Plasma noradrenaline (NA) levels at slaughter appeared to correlate closely with the observed excitability of groups with control space allocation and also with the aggression scores of the three groups.

TABLE 1

EFFECT OF STOCKING RATE ON THE BLOOD PROFILE OF LARGE WHITE BOARS

Treatment			Responses in pens		Responses at slaughter		
	Group size	space /pig (m^2/pig)	Aggression score/ h/pig[a]	Plasma cortisol[b] (µg/100 ml)	Plasma cortisol[c] (µg/100 ml)	Plasma cortisol[c] (ng/ml)	Plasma noradrenal (ng/ml)
Large restricted	11	0.39	2.2	2.5	4.4	29.7	19.4
Large control	10	0.88	2.6	2.6	3.5	33.7	27.3
Small control	5	0.86	3.7	2.9	4.5	32.1	35.5

[a] The aggression score was calculated by multiplying the number of fights by their appropriate scores and expressing these for each hour of observation/pig in the group.

[b] Plasma cortisol responses in the pens were determined on 0.1 ml blood samples obtained from a needle prick on the ear after restraining the pig with a noose.

[c] Responses at slaughter were measured by determining the levels of the various parameters in blood collected at slaughter.

This effect of adaptation in the responses of blood parameters to given stressors can be examined further by studying the cortisol and thyroxine levels in pigs of different breeds. It is generally accepted, on the basis of meat quality, that Landrace pigs (LR) are more stress sensitive than Large White pigs (LW) (Bendall et al., 1966). Piétrain pigs have been observed to have greater catecholamine responses to stressors than Large White pigs (Hall et al., 1975). Similarly, Landrace pigs are expected to be more excitable and exhibit hormonal responses typical of stress sensitive breeds. If cortisol levels are related to the level of stressor (Baldwin and Stephens, 1973), then the smaller increase in plasma cortisol levels in Landrace pigs than in Large White pigs, due to pre-slaughter handling (Table 2) is paradoxical, considering the greater stress sensitivity and poorer meat quality of the Landrace breed. Similarly, the thyroxine response may indicate that LR pigs are so sensitive that a maximal thyroxine response (an increased utilisation of thyroxine) (Moss, 1975) occurs even when subjected to the stressors of handling and restraint. It has been suggested that in stress sensitive breeds of pigs, the adrenal response to ACTH is lower and the rate of cortisol utilisation higher, than in stress resistant pigs (Marple and Cassens, 1973). Accordingly cortisol responses would be expected to follow a type 2a curve for Large White pigs and a type 2b curve for Landrace pigs. This is supported by the experimental results of Table 2, although it contradicts the responses to pharmacological stimulation of the hypothalamic-pituitary-adrenal axis reported by Lister et al. (1972).

In experiments where parameters of both adaptive and non adaptive responses in the blood profile were measured (Moss and Robb, 1978; Moss and McMurray, 1979), there is evidence to suggest that serum CPK and LDH activities can be used to interpret the amount of stressor the pigs were subjected to in the previous 24 h. In both boars and gilts CPK levels at slaughter were higher after 24 h in lairage than 2 h in lairage and the increase due to lairage was significantly greater in boars than gilts (Table 3). This sex difference possibly reflects the greater

TABLE 2

CORTISOL AND THYROXINE RESPONSES OF LANDRACE AND LARGE WHITE PIGS TO THE STRESSORS OF RESTRAINT AND PRE-SLAUGHTER HANDLING[a]

	Plasma cortisol (μg/100 ml)		Plasma thyroxine (μg/100 ml)		
Breed	in pens[b]	at slaughter[c]	in pens[b]	at slaughter[c]	
Landrace	2.3	4.6 (***)	4.0	4.2	(NS)
Large White	2.8	6.0 (***)	4.9	4.1	(*)

[a] The pigs from their individual pens were mixed on the lorry, transported for a period of 35 min to the abattoir and killed by electrical stunning after 1 h in lairage.

[b] The pigs were restrained by a noose round the upper jaw, and blood samples were collected from an ear vein within 5 min of restraining the pig.

[c] Slaughter samples obtained at exsanguination.

TABLE 3

EFFECT OF TIME BETWEEN FARM AND SLAUGHTER ON SOME PARAMETERS IN THE BLOOD AT SLAUGHTER

		Blood profile at slaughter					Meat quality	
				Serum enzymes				
Sex	Time[a] to slaughter	Plasma cortisol	Plasma thyroxine	CPK	LDH	GOT	PSE[b]	DFD[c]
	h	μg/100 ml	μg/100 ml	iu/l	iu/l	iu/l	%	%
Boars	2	6.3	4.1	1696	704	63	22	19
Gilts	2	6.4	4.6	3095	816	71	59	4
Boars	24	5.5	4.7	15353	1102	140	12	81
Gilts	24	5.0	5.7	8627	1101	105	25	31
Significance of:								
Sex		NS	*	NS	NS	NS	*	***
Time to slaughter		**	**	***	NS	***	**	***

* $p < 0.05$; ** $p < 0.01$; *** $p < 0.001$; NS = not significant

a - duration of transport journey accounts for 15 min of this period, the remainder is time in lairage

b - *longissimus dorsi* muscles with pH values of 5.9 or less, 45 min *post mortem*

c - adductor muscles with pH values of 6.2 or greater, 24 h *post mortem*

CPK - creatine phosphokinase; LDH - lactate dehydrogenase; GOT - glutamic oxaloacetic transaminase aggressive activity of boars than gilts when held in lairage for long periods. This suggestion is further collaborated by the higher incidence of DFD after 24 h in lairage (Moss, 1978). Hormonal parameters at slaughter would tend to indicate that after 2 h lairage the pigs were more 'stressed' or more sensitive to the stressor of slaughter than after 24 h in lairage. The decrease in corticosteroid level at slaughter observed from the 2 h to 24 h period would be predicted from the model system used for Figure 2.

6. CONCLUSIONS

From the few examples quoted it would appear that blood profiles have an important role to play in the assessment of stress, particularly if several parameters are measured. There is, however, a place for behaviour studies, as they can provide confirmation of the blood profile results.

REFERENCES

Baldwin, B.A. and Stephens, D.B., 1973. Physiol. and Behav., 10, 267-274.

Bareham, J.R., 1973. Vet. Rec., 92, (25) p682.

Bendall, J.R., Cuthbertson, A. and Gatherum, D.P., 1966. Proc. XI Meeting European Meat Research Workers, Belgrade.

Ewbank, R., 1973. Vet. Rec., 26, 709-710.

Fraser, A.R., 1973. Vet. Rec., 93, 27.

Hall, G.M., Lucke, J.N. and Lister, D., 1975. Anasthesia 30, 308-317.

Lister, D., Lucke, J.N. and Perry, B.N., 1972. J. Endoc., 53, 505-506.

Marple, D.N. and Cassens, R.G., 1973. J. Anim. Sci., 36, 1139-1142.

Moss, B.W., 1975. PhD thesis Univ. of Bristol.

Moss, B.W., 1978. Appl. Anim. Ethol., 4, 323-339.

Moss, B.W. and McMurray, C.H., 1979. Res. Vet. Sci., 26, 1-6.

Moss, B.W. and Robb, J.D., 1978. J. Sci. Fd. Agric., 29, 689-696.

Napier, J., 1973. Br. Vet. J., 130, 85-86.

Sebranek, J.G., Marple, D.N., Cassens, R.G., Briskey, E.J. and Kastenschmidt, L.L., 1973. J. Anim. Sci., 36, 41-45.

Selye, H., 1951. Stress, Acta Endoc. Inc., New York.

DISCUSSION

V. Tarrant *(Ireland)*

In the last slide the figures for hormones and serum enzymes indicated that the boars were more stress-susceptible than the gilts in general. But the figures for PSE meat indicated that this condition was higher in gilts than in boars. How would you interpret this? We would normally associate PSE meat with higher stress sensitivity in the pig.

B.W. Moss *(UK)*

If we look at the incidence of PSE it is greater in gilts than in boars. There is a problem in this in that if we have a pig that is highly sensitive, from my observations it will be more excitable. This may not apply to all PSE pigs. However, if we have a pig that is stress sensitive and excitable and then subject it to stress, it may result in more physical activity which will deplete the glycogen level to a certain extent. This will mean it is on a borderline between PSE and normal. I have then seen pH1 values of 5.9 and 6.0. I do not think these would be the typical pale meat PSE conditions. I believe that if you could look at a very short duration stress (which is difficult because of previous stress sensitivity) then this would be reversed and there would be a higher incidence of PSE. It is a problem that they switch over from a PSE to a DFD condition due to the depletion of the glycogen level already mentioned.

R. Dantzer *(France)*

Large White, which are described as stress non-susceptible animals, when they are stressed during transportation tend to make dark, firm dry meat instead of having PSE.

First, when you talk of adaptive response versus non-adaptive response, I do not think that this is very useful since it could cause misjudgement about the patient.

You describe this type of response in terms of kinetics, but I think this is sufficient and that you do not need to talk about adaptive or non-adaptive responses.

When considering enzyme systmes, such as CPK and ADH you have to keep in mind that these enzyme systems only react when the animal shows physical activity. So, if you put the animal in a strictly psychological situation, such as exposure to frustration or a new environment in which there is little physical activity of the animal, you would not expect any change in the enzyme activity. Do you agree?

B.W. Moss

Yes, and in fact I would take that point further. I intended to make the point with the differences in serum enzymes in the slide. I am taking the normal interpretation that corticosteroids increase in stress, yet the lairage itself may be a stress and I have observed that there is quite a considerable amount of aggression and physical activity, particularly in boars when they are penned in lairage overnight. That is why I was suggesting that serum enzymes can be used. Obviously the interpretation that you have put up is that in a stress situation we measure the three parameters that I have suggested - some serum enzymes (you want a specific one for the particular type of stress), some measure of emotional stress and some measure of the stress of actual sampling - that if there are low serum enzyme levels, this does not necessarily mean there is low stress, it means there is low physical type of stress and obviously there may be more emotional stress. We know that serum enzymes are a measure of two things - physical stress and the sensitivity of the membrane, which may be different between breeds and sexes. We can use serum enzymes to tell us something about the aggressive activity and the aggressive activity within the previous 12 or 24 h. Corticosteroids would be limited to use maybe 1 or 2 h after the stress. I would agree with your point, and would emphasise that it is a question of the balance of the three parameters that I suggest should be used in interpreting the previous stress history of the animal and maybe helping to support behavioural observations.

R. Dantzer

My third point is not for practical application in which you are dealing with several stressors acting at different times rather than all at once. We are lacking much fundamental data which are related to such concepts as tolerance or sensitisation. In other words, we do not know yet if pigs are able to show tolerance to a stressor and cross-tolerance to other stressors, or whether they show sensitisation which is an increased response after a previous exposure to a stressor. I think that this is a very important limitation for the interpretation of present data.

B.W. Moss

I think that in Table 1 the stocking density shows evidence that there can be adaptation and certainly this can be seen in terms of excitability. If one makes a noise in the pens the pigs that have most room will run around and this will lead to aggressive activity. At the same time, the confined animals with less space allocation cannot run around and therefore they are deprived of a normal behavioural trait. However, is it an advantage for the pigs to run around when this results in greater aggressive activity and greater stress? In terms of meat quality, the group that had a large space allocation and a small group size had the worst meat quality. The other two groups were not markedly different in this respect. So there is obviously some adaptation in their sensitivity to stress and different stress responses. Probably several hormonal systems are involved.

D.L. Ingram *(UK)*

I would like to ask you about use of plasma thyroxine levels. You mentioned levels and also utilisation rates.

Were you doing both, or inferring one from the other?

B.W. Moss

I mentioned utilisation rates because, as I predicted from the first curves, if you actually measure a parameter in the blood, it is a balance between secretion and removal. It may be that under a stress situation both may change, but in certain breeds this may result in the increase of utilisation relative to secretion. This could explain some of the difficulties in interpreting stress sensitivity, particularly from the thyroid point of view - where high thyroxine levels are related to increased excitability.

G. van Putten *(Netherlands)*

When I first read these two papers, I thought there was some margin of disagreement between the two authors; Dr. Dantzer who told us of the need for care in using blood profiles in order to assess well-being in pigs, and Dr. Moss who relied quite heavily on these blood profiles. However, from the discussion I gathered that the differences are not as wide as I first thought, and yet I wonder if you could comment on the differences between the two papers, especially in regard to long-term stress.

B.W. Moss

In my paper, I proposed a method by which we may be careful in interpreting blood profiles. Obviously there may be other ways, but this is the way that I propose.

In terms of adaptation, I believe that this is a major problem, and I think that only with extensive research can we begin to balance out the behavioural side and the physiological side. I do think that we have to concentrate on the actual practical situation. In terms of the blood profile, limits may be set for plasma corticosteroids and possibly for behaviour as well. We must try to balance the two. I would suggest that if we are going to use behaviour to assess stress, we need experts to go round different units. If we use a blood profile then any qualified person, veterinary practitioner or technician, could actually take the sample, provided that the right observations were made at the time of sampling, and then only one expert would be needed to interpret the results. I do agree with Dr. Dantzer that we cannot rely on one hormonal measurement or one serum enzyme measurement; I suggest that we need to consider several parameters, some of which may tell us something about long-term adaptation. I cannot say which is the best to use now since I believe that corticosteroid responses change, certainly in relation to stocking rate and group size.

W. Sybesma *(Netherlands)*

I do not agree with your observation that stress-susceptible pigs are in fact stress-susceptible. If we talk about stress-susceptible in relation to the cortico hypophysis system, then that is different from talking of stress-susceptible pigs in relation to PSE and DFD. When there is a breakdown of the muscle system it does not mean there is also a breakdown of the hypophysis adrenal system. You might say that it is generally accepted that PSE is the same as stress-susceptible with regard to hypophysis adrenal system, but I think that they are two different systems. I think that you should make a division between these two systems, and when talking about stress we should be clear about what is actually understood by it.

Secondly, I would like to remark that the very high concentration of CPK could also be due to tissue damage. So, for instance, if there is bleeding or a contusion there will be a high concentration of CPK which has little to do with aggression or stress.

B.W. Moss

I think that maybe you have misunderstood my first point. I was showing the relative differences within breeds in terms of response. I am not saying that pituitary, adrenal response is responsible for the differences in stress sensitivity to the pig. It would be accepted that Landrace and Large White pigs reveal differences in stress susceptibility, as indicated by meat quality. Along with this there are differences in adrenal corticosteroids, and in thyroxine. However, there are also differences in the way that these pigs respond to the stressors of mixing and transport. It is not to say that thyroid function of adrenal corticosteroids is the answer.

In terms of membrane integrity, I would agree with you that high CPK levels could indicate blood damage. However, I would ask how that blood damage arises? What are the reasons for the high levels after 24 h lairage? I would suggest bruising resulting from fighting - that is supported by behavioural observations that boars fight more in overnight lairage than gilts. So I agree that there is a problem in interpretation, but that if we stick to the detail and say that serum enzymes indicate that there is a physical breakdown in membrane integrity, we are going to get no further in interpreting stress. So, I must stress that it is the problem of interpretation of all the various parameters that the behaviourists and the physiologists cite and we must think more about 'why' rather than 'this happens'.

GENERAL DISCUSSION

J.P. Signoret *(France)*

In order to orientate this discussion, I propose three main subjects for consideration:

1. Behavioural needs: the capacity of adaptation and versatility of the pig: the importance of genetic variations, etc.

2. Criteria: the validity of ethological criteria, of normal behaviour. Does this involve some adaptive components, as Dr. Dantzer's results suggested, or should this be taken as an absolute criterion for welfare? The validity of physiological criteria; and the validity of preference - when allowing the animal to choose - does this reveal simple preferences or a true welfare situation?

3. What are the possibilities in connection with environment and management practices for changes in the welfare situation?

D. Smidt *(FRG)*

Papers have been given not only on needs but also on concepts: how to assess these needs and how to assess whether these needs are met or not in animal husbandry. This leads to the question of welfare indicators, and I think that it is quite clear that we must accept the necessity of a multiple system of welfare indicators - ethological, physiological and even medical parameters. I would like to draw attention to a problem that has been discussed in all the welfare meetings that I have taken part in so far; whether we can draw the conclusion that if ethological changes or disturbances occur under certain management systems, that these systems are not in accordance with the needs of the animals. In Germany this is a very serious question because the Animal Welfare Act of 1972 stated that animals must not be kept in an environment that is not in accordance with their needs. Should we, therefore, not put these categories of indicators side by side, but rather to try to interpret criteria in one category by findings in other categories. In terms of research in this area perhaps it would be useful to correlate ethological criteria with physiological criteria so that in those instances when we are not sure whether the animal suffers or not, we can have a better chance of interpreting these ethological situations?

R. Dantzer *(France)*

If we take, just as an example, the model shown in my paper concerning displacement activity in pigs permitted intermittent food delivery, they developed very high sensitivity, and at the same time they very effectively controlled their plasma corticosteroid levels. This leads to the question of what can be done in practice. Dr. Vestergaard had described this sort of behaviour as abnormal, but it can be seen as normal behaviour if its consequences are seen in terms of hormonal in the sense that it enables the animal to keep homeostasis. To answer the question we must look at what is occurring at other times. If blood samples are drawn from sows that are showing bar gnawing or any other repetitive behaviour, I am sure that you will find that these sows have normal corticosteroid levels. This has been reported by Duncan and Wood-Gush in poultry. However, there is still something abnormal in this hormonal response; before decreasing their plasma corticosteroid levels the animals have to increase them. There is some build-up of what might be called anxiety or tension. The hormonal level just before the beginning of the session is higher than normal, but by the end of the session the level is normal - this is an important point. The animal has adapted behaviour enabling it to regain homeostasis, to go from an abnormal high level to a normal, low corticosteroid level. This cannot be explained unless both the behavioural response and the endocrine response are examined.

J.P. Signoret

When thinking of the process of domestication, we must consider that when the first human beings domesticated animals they created a very special ecological niche. Animals had to be kept under their control which means they had to tolerate either being tethered or kept under relatively close confinement. Therefore, it is not surprising to find the huge adaptive possibilities resulting from what Dr. Mauget has told us. This adaptability of domestic species involves everything: space and time use, food use, and also the tolerance to suppression of some patterns of behaviour. I think that this is open for discussion. There may be factors that make a given animal more suitable for domestication. For instance, the absence of a strong male-female bond - if this had existed in domestic species great difficulties would have been encountered in practical breeding. The same applies to species which exhibit a strong maternal bond. We must consider that the animals are not only exhibiting all the natural patterns of behaviour, but also what they can tolerate in the suppression of their natural behaviour. What is the level of adaptability from an abnormal behaviour, and what is the possibility for the animal to return to a normal physiological function, even when a number of environmental circumstances have changed?

R. Moss *(UK)*

We have looked at a number of parameters and then said implicitly that these parameters can equate with good or bad welfare of the animals concerned. We have looked at the economics, the stress factors, changes in behaviour; we have looked from the outside in; but we have had very few attempts apart from Dr. Ingram's paper, to look from the inside out. I realise that it is very difficult to find out what a pig, or any animal, is perceiving, but why has there been so little mention of, or even an attempt at experimentation in, this area?

G. van Putten *(Netherlands)*

I think that we are dealing with two 'independent' branches of science - physiology and ethology. Neither is dependent on the other. They can each add to the understanding of the other, but I must stress that they are independent, and this is an important point for us to agree upon.

In ethology, I believe that we should not concentrate on the traits of behaviour an animal is not showing, compared with an animal in a more natural environment; I think that we should try to find out whether the animal kept in modern intensive husbandry systems can cope with the situation as it is. We must ask whether there are types of behaviour like conflict, vacuum behaviour, stereotypical behaviour etc., which show us that the system of housing and husbandry lies beyond the adaptive capacity of our domestic animal.

V. Fischbach *(Luxembourg)*

My question is related to a welfare indicator that has not been mentioned so far. I was wondering why physiological behaviour during sleeping time has not been studied so far? It is known that human beings, under stress, do not have REM cycles (Rapid Eye Movements) during sleep. Experiments carried out in the UK reveal the same with race horses. Horses transported only one day prior to a race had not REM periods when they slept. Performances afterwards were low to very low.

So, when pigs suffer stress, do they have REM periods or disturbed REM cycles? If they have these periods, are the pigs feeling well? I believe REM periods may be a good parameter for studying the welfare of pigs.

D.L. Ingram *(UK)*

I would like to raise a point that might well seem heretical. We have discussed welfare from the point of view of the pig, however, the point is that welfare is really something that arises because of people objecting to the way in which animals are kept. We hear they object in Germany and they certainly object in Britain. It seems to me that one of the criteria that we must meet is that the animals must be kept in a manner that does not evoke this sort of response from the general public. Whether the pigs are happy or not; if people do not think that they are happy, the system will be rejected.

M. Zanforlin *(Italy)*

Pigs are very expensive animals to experiment with and this may be why ethologists do very little work with them. We have little knowledge about how important the environment is from the point of view of the pig, but being inquisitive animals they probably need a more exciting environment giving more stimulation which may decrease some types of stereotypical behaviour.

R.G. Buré *(Netherlands)*

It may be possible to make links between physiological and ethological research when it is done immediately after a change from one housing system to another. Perhaps change is very important for the animals and straight away after the change it may be possible to combine physiological and ethological research. After a period of, say, 3 or 4 weeks, the animal can adjust to a certain housing system by making some behavioural adjustments. Abnormal behaviour could be an attempt to manage in a housing system such as a full slatted floor or a battery cage. If the animal manages, the physiological situation may return to normal, but ethological observations may still reveal abnormal behaviour. As a conclusion, I would say that physiological and ethological research should be compared as soon as possible after a change in environment.

J.P. Signoret

Are you proposing that ethological criteria could be more sensitive than physiological ones?

R.G. Buré

No, physiological and ethological combinations are necessary. Internal responses can be revealed in behaviour, so sometimes a physiological response is due to an ethological reaction, and also the reverse can be true. The two disciplines can help each other in such a case. But when talking about welfare, a division must be made; sometimes it is possible to combine the two sciences and sometimes not.

<u>B.W. Moss</u>

In relation to the last comment and a previous comment by Dr. Dantzer, I wonder if we should stop measuring behaviour and blood profiles, since I believe that we are just accumulating data.

The interpretation is important. An abnormal behaviour can be an advantage for the animal if it moves away from the stressful situation. Therefore, it is our interpretation which counts!

<u>D.L. Ingram</u>

We keep talking about 'advantage' and 'is it better'. Biologically, if an animal survives so that its DNA gets into the next generation, it has been a success. It does not matter biologically whether the animal is happy or not. Welfare is a human criterion; the animal will adapt using physiological, behavioural mechanisms to stay within certain limits, thus assuring the next generation.

<u>J.P. Signoret</u>

However, when we are dealing with fattening animals, we cannot be sure that they will be able to transmit their DNA to the next generation.

<u>Doris Buchenauer</u> *(FRG)*

I do not think that the animals, in this or the next generation, will adapt - we manipulate them too much for this. To keep the animal alive medically does not mean that the animal has adapted. We cannot look at this question in the way that Dr. Ingram suggests.

<u>T. Hanrahan</u> *(Ireland)*

I agree with Dr. Ingram but would like to broaden what he said. I agree that welfare is a human word and a concept that is hard to define.

However, there is another area that brings in physiology. There is pressure to produce meat efficiently. There are many parameters depending upon the subject that is being discussed, and there are two areas that we must follow: first, the physiological and production side, and welfare as the second criterion.

SESSION III

COMPARATIVE DATA REGARDING WELFARE IN PRESENT HUSBANDRY SYSTEMS, I.

I. FATTENING PIGS

Chairman: R. Moss

OBSERVATIONS ON THE EFFECTS OF STOCKING RATE ON THE PERFORMANCE OF GILTS AND BOARS TO BACON WEIGHT

T.J. Hanrahan
Moorepark Research Centre, Fermoy, Co. Cork, Ireland.

1. INTRODUCTION

Behaviour of pigs in response to their environment is both of acedemic and practical interest. The scientist may desire to know the physiological and biochemical explanation for responses or he may wish to study productivity, labour efficiency or profitability in animal production. The development of large scale production units has added a further area to be evaluated namely, the welfare of stockmen and stock. This however, is an area that requires further definition and as yet, the criteria which we use may not even be adequate. In all probability, a concern for the welfare of men and stock and an interest in profitable production are not mutually exclusive.

The construction of a new relatively large pig fattening house at Moorepark Research Centre raised a number of questions, such as what was the optimum stocking rate or what feeding system would allow experimentation without placing undue demands on labour. There is a considerable literature on the floor area required by pigs (Jensen et al.,1963; Gehlbach et al., 1966; Heitman et al., 1961; Plumlee et al., 1976; Bryant and Ewbank,1973). However, many of the experiments reported in these publications deal with small animal groups, a wide variation in space allowance and indicate a reaction between season and stocking. Furthermore, many experiments relate to *ad libitum* feeding systems which may not be economical in all pig producing countries (Braude, 1971).

The experiment reported in this paper was designed to evaluate the effect of two stocking rates on the performance of boars and gilts. The feeding system used allowed some measure of feed restriction.

2. EXPERIMENTAL PROCEDURE

2.1. Fattening house

The house measures 48.8 x 11.25 m internally and has 40 pens in a common air space each measuring 4.7 x 2.3 m. Pens are on either side of a centre passage and floors are partially slatted (25% slat). Water is available *ad libitum* from a drinking nipple sited over the slat and feed is available from a hopper type feeder (1.5 m feeding space) sited on a dividing wall. Ventilation is natural, there is a continuous controlled air inlet along both sides of the house (0.17 m in width) and 10 chimneys sited along the centre passage (each 0.6 m^2). The floors, walls and ceiling are insulated. Throughout the duration of the experiment, temperature (max - min over 24 h) was recorded daily from 6 thermometers spaced through the house. Air inlets and outlets were adjusted twice daily if necessary to maintain an even temperature within the house.

2.2. Pigs

The experiment involved 1 360 pigs which were penned in groups of 16 and 18 on the basis of sex and initial weight. These stocking rates allowed a floor space per pig inclusive of slat but exclusive of feeder, of 0.64 and 0.56 m2 respectively. The experimental period was from 30 to 80 kg liveweight. As pigs approached 80 kg they were weighed weekly and those exceeding this weight were sent for slaughter. When the fourth last pig in a pen exceeded 82 kg the remaining three pigs were slaughtered irrespective of their weight.

2.3. Feeding

A fortified barley-soyabean meal ration, containing 12.6 MJ DE/kg and 0.8% lysine was fed throughout the experiment. Feed was delivered to the piggery in pre-weighed bags and the quantity allocated to each pen was based on the feed scale on Table 1, or was sufficient to allow a 6 hour feeding period. Feeding was at 9 am daily and if a pen had feed available for more than 6 hours on any particular day, the meal allocation

TABLE 1
FEED SCALE

Pig weight kg	30	40	50	60	70	80
Feed/day kg	1.4	1.64	2.0	2.26	2.5	2.76

was reduced the following day.

2.4. Results

House temperature is shown in Table 2. There was little variation in temperature with season. In fact the day to night temperature change ca. 5°C was greater. This fluctuation would not have a significant effect on performance (Close and Mount, 1978).

TABLE 2
AVERAGE MONTHLY TEMPERATURE °C DURING EXPERIMENT

Month	Average Maximum	Average Minimum
July 1979	26.2	21.9
August	26	20.7
September	26	20.6
October	25	20.4
November	22.9	18.6
December	22.2	18.6
January 1980	23.4	17.6
February	24.2	19.1
March	25.1	20.8
April	24.3	19.0
Average SD monthly mean	± 1.4	± 1.6

The performance of pigs outlined in Table 3 is basically good. Of 1360 pigs allocated, 5 died or were taken off the experiment. The higher stocking rate resulted in a significant deterioration in feed intake ($P < .01$) daily gain ($P < .001$) and feed utilisation ($P < .05$). Boars grew faster and converted their feed more efficiently than gilts ($P < .001$).

TABLE 3

EFFECT OF STOCKING RATE AND SEX ON PIG PERFORMANCE

Stocking rate m^2/pig	0.64		0.57		SE	Sig.	
	Male	Female	Male	Female	of mean	SR	Sex
Number of pigs	320	320	360	360			
Initial weight kg	32.5	32.5	31.5	32.2			
Final weight kg	83.3	83.2	83.3	82.3			
Feed intake/day kg	2.13	2.16	2.08	2.09	±0.017	**	NS
Gain/day g	716	656	680	617	± 5	***	***
Feed/kg gain kg	2.99	3.30	3.06	3.40	± 0.02	*	***

Pigs on the high stocking rate and gilts had a higher kill-out percentage ($P < .01$) (Table 4). Boars had the leaner carcasses ($P < .01$) while stocking rate had no effect on carcass quality. There were more underweight pigs on the high stocking rate indicating greater variation in weight at the end of the experiment and while this has been reported as a feature of high stocking rates (Ford and Teague, 1978) the nature of the present data where the effect is confined to gilts prevents any conclusion. It may however, be an area that merits further research as there are reports in the literature suggesting that gilts have a greater space requirement than males (Hugh and Reimer, 1967; Jensen, 1971).

Finally, the effect of stocking rate on veterinary interventions and mortality is outlined in Table 5. The number of injections was similar for both treatments which incidently need not imply a similar percentage of pigs, as repeated injections of the same pig were not recorded. Mortality was small and followed no pattern.

TABLE 4

EFFECT OF STOCKING RATE AND SEX ON KILL-OUT PERCENTAGE AND CARCASS QUALITY

Stocking rate m^2/pig	0.64		0.57		SE of	Sig.	
	Male	Female	Male	Female	mean (2)	SR	Sex
Kill-out percentage	70.9	72.5	71.4	73.1	±0.11	**	**
Average backfat mm	25.9	27.4	25.9	27.1	±0.1	NS	**
Pigs under 50 kg DW(%)	0.9	0.9	1.7	2.2			

TABLE 5

EFFECT OF STOCKING RATE ON VETERINARY INTERVENTIONS AND MORTALITY

Stocking rate m^2/pig	0.64	0.57
Pig injections (as percentage of pigs allocated)	25	26
Pig injections (actual number)	161	188
Pig mortality number	3 (1 sudden, 2 chronic)	1 (1 sudden)
Pigs removed number	0	1 (attacked)

3. DISCUSSION

The effect of stocking rate observed in this experiment is basically in agreement with previous publications. Gehlbach et al. (1966) found that a floor space of 0.36 m^2 and 0.54 m^2 per pig from 50 to 70 kg and from 70 to 90 kg respectively, significantly reduced growth rate but had no effect on feed efficiency. Daily feed intake was reduced which these authors suggest is the first noticeable response of pigs under the stress of crowding. Hugh and Reimer (1967) also reported that daily gain but not feed efficiency was significantly affected by a floor space of 0.28 and 0.56 m^2 for pigs 16 to 45 kg and 45 to 77 kg respectively. The present study however, differs from both of these studies in that pigs were not fed *ad libitum*. The greater number of pigs in the higher stocked pens had a smaller feeding space per pig 8.3 cm versus 9.4 cm. This resulted in considerably more aggression which persisted for a longer period at feeding time. Experiments where more severe space restriction was practised e.g. 0.43 m^2 per pig, reported a significant deterioration in both growth rate and feed efficiency (Plumlee et al., 1976; Krider et al., 1975).

The effect of stocking rate on kill-out percentage simply reflects the lower feed intake of the higher stocked pigs, while the effect of sex on pig performance and carcass quality are in agreement with previous studies (Walker, 1977 - 78).

CONCLUSION

In conclusion, the effects of higher stocking rate observed in this experiment were confined to reduce feed intake and poorer feed efficiency. There was no apparent effect on the health or comfort of pigs with the exception of increased aggression at feeding time in the higher stocked pens. Yet, the deterioration in feed efficiency indicates considerable stress as there are many examples in the literature (Braude, 1971) and it is also the experience of this author that a reduction in feed intake of the magnitude observed on the higher stocking rate should in fact result in an improvement in feed efficiency.

REFERENCES

Braude, R., 1971. Pig Production Univ. of Nottingham 18th Easter School in Agric. Sci.

Bryant, M.J. and Ewbank, R., 1973. Br. Vet. J. 130, 139.

Close, W.H. and Mount, L.E., 1978. Br. J. Nutr. 40, 413.

Ford, J.J., Teague, H.S., 1978. J. Anim. Sci. 47, 828.

Gehlbach, G.D., Becker, D.E., Cox, J.L., Harmon, B.G. and Jensen, A.H., 1966. J. Anim. Sci. 25, 386.

Heitman, H. Jr., Hahn, L., Kelly, C.F. and Bond, T.E., 1961. J. Anim. Sci. 20, 543.

Hugh, W.I. and Reimer, D., 1967. J. Anim. Sci. 26, 891. Abs. 37.

Jensen, A.H., 1971. Unpublished data cited J. Anim. Sci. 32, 560.

Jensen, A.H., Becker, D.E. and Harmon, B.G., 1963, Ill. Agr. Exp. Sta. Publ. A.S. - 587.

Krider, J.L. et al., 1975. J. Anim. Sci. 40, 1027.

Plumlee, M.P., Cline, T.R., Krider, J.L. and Underwood, L., 1976. J. Anim. Sci. 42, 1339, Abs. 9.

Walker, N., 1977-78. 51st Ann. Rep. Agric. Res. Inst. Nth. Ir.

DISCUSSION

R. Dantzer *(France)*

One of the main effects of competition for food is to increase variation between individual animals in terms of weight gain. Did you observe, in Table 2, that there is not an indication of standard error in the variation of each group, or was the variance homogeneous between the groups?

T. Hanrahan *(Ireland)*

There is only a very small variation. The percentage of pigs slaughtered at a dead weight of under 50 kg, was 0.9 at the lower stocking rate. When you think of the large number of pigs involved it is not as significant as one might expect.

I. Ekesbo *(Sweden)*

You observed the effects of higher stocking rates. Are you sure that the effects that you saw are due to a higher stocking rate and not to some extent due to the area per pig at the feeding trough? We know that there is much more aggression within groups where there are conflict situations during feeding periods. Our studies indicate that the situation during feeding-time seems to be more important than the stocking density as to the aggression situation within the group.

T. Hanrahan

In my observations, there was more aggression at feeding-time with the higher stocking than with lower stocking. Feed was available for six hours for all groups, and periods existed when there were no pigs feeding even though food was available. I am not overlooking the effect of smaller trough space per animal in that it did lead to more aggression at feeding-time, although it did not develop into vice aggression habits such as fighting and tail-biting. I do believe that we can overstock pigs and cause stress that is difficult to measure in terms of vice or aggression. One should look at the physiological implications of a reduction in feed conversion efficiency by pigs that are under stress.

K. Vestergaard *(Denmark)*

I would like to comment on the amount of feeding-space available. A well-controlled Danish study showed that even the *ad libitum* feeding using five feeding automats instead of one, gave a reduction in the amount of aggression with the same space allocations to the animals. I think that in your experiments it was a combined effect of a reduction in space per animal and a reduction in feeding-space.

One more point, how do you know that tail-biting is aggression?

T. Hanrahan

One cannot definitely know this, but one can find many references in the literature to tail-biting and fighting as being criteria of overcrowding.

K. Vestergaard

The trouble is that nobody has studied the motivation of tail-biting. In ethological terms, a motivation analysis has never been carried out, probably because it is so difficult to observe tail-biting and the circumstances in which it occurs.

T. Hanrahan

Again, I think this would be borne out by my experiments where there are practically 700 pigs per treatment. There is also very little difference in the number of veterinary interventions, in the number of mortalities (although mortality was even greater in the lower stocked pens). So, I believe, as maybe you do, that these are not the best indicators.

W.T. Truyen *(Netherlands)*

You have been looking at the number of pigs/m^2 but surely it is also important to look at the number of pigs in the group. The size of the group can affect the stress that occurs at feeding-time, and this morning we heard about the organisation of the group, the hierarchy, which I think is an important point in relation to this question.

T. Hanrahan

Yes, I think that it is important and I believe that it is well documented in the literature. As group size gets larger there is more stress on animals, and the physical performance as measured by the normal production criteria deteriorates. At our institute we have individual feeding facilities for pigs right through to group sizes of five, ten and sixteen to eighteen which is the largest. If we look back over any period on the performance of our animals we find better performance at the smaller group size. However, there is a conflict when one discusses this in relation to practical production systems in that it is expensive. We are looking for a compromise and so find ourselves with a variety of pen sizes.

W.T. Truyen

The difficulty is that we are talking about welfare and not about economics.

T. Hanrahan

We have recently taken a pig house from a commercial manufacturer where pigs are penned individually, and already welfarists are on to us claiming that this is not a suitable house because it is divided in cages. They would like individually penned animals and yet there are certain types of individual penning that they do not believe are acceptable.

J. Daelemans *(Belgium)*

You stated, Dr. Hanrahan, that all European countries practise restricted feeding. In Belgium, however, we are an exception; as a rule we have *ad libitum* feeding. We have Piétrain and Belgian Landrace breeds. The consumer demands high meat quality with ham and cutlets. There is a wide variance in prices, between high and normal quality, and this is a different situation from that found in most other countries. Our pig farmers have been stimulated by consumer demand to produce a high meat quality pig, which is one reason why we have a different breed.

CHARACTERISTICS OF PIG FATTENING SYSTEMS INVOLVING PARTIALLY SLATTED FLOORS

A.T. Smith
Agricultural Development and Advisory Service,
Wolverhampton, UK.

1. INTRODUCTION

The impact of production methods on the well-being of the pig is governed by the interaction of all the component parts of the system. Thus, design, construction, installation and management together with day to day stockmanship, all have an influence on the welfare of the animal. There is no aspect of pig production in which this is more clearly evident than in systems of pig fattening which involve the use of partially slatted floors. This is because the system allows the pigs to give a clear behavioural response to changing environmental conditions. Such behavioural changes are not necessarily related to performance parameters but they usually give the stockman an unmistakeable signal in the form of dirty pens. In this paper I wish to focus attention on four aspects of such systems:

a. The aim of the system and an indication of its importance in the United Kingdom together with a description of some of the variations found in commercial use.

b. An examination of the factors which influence the operation of fattening pens involving partially slatted floors.

c. A summary of the characteristics of the system.

d. An assessment of the extent to which partially slatted floors achieve their objective.

2. AIM, IMPORTANCE AND DESCRIPTION

2.1. Aim

The primary objective in incorporating slatted areas into pen floor design is to minimise labour requirements.

A secondary objective, in non-cereal growing parts of the country, is to eliminate the cost of bedding.

2.2. Importance

There are no national statistics on the proportions of pigs housed under different systems but an analysis of the production methods used by 490 members of the Meat and Livestock Commission's recording schemes (1979) in the United Kingdom showed 42% have fattening accommodation incorporating partially slatted floors. In contrast only 1.4% of herds have fully slatted fattening pens. This is not a random sample and the national proportion of pigs housed on partially slatted floors may be somewhat lower. It is also likely that the figures fail to reflect the strong current interest in fully slatted floors for fattening pigs.

In the same sample an analysis of pig finishing systems involving all types of floors revealed the following general housing classifications:

Completely closed and insulated	- 39%
Kennel type	- 27%
Verandah	- 9%
Open fronted mono-pitch	- 10%
Other types and combinations	- 15%

There is no detailed break-down of the proportion of each classification which uses partially slatted floors but the analysis is sufficient to indicate that a wide variety of systems are in common use and that kennel, open fronted, and verandah types make up a substantial proportion of the total.

2.3. Description

The four types (Appendix 1) described are:

i. Enclosed, insulated building with fan ventilation
ii. Open-fronted mono-pitch
iii. Kennel verandah.

iv. Open, uninsulated building incorporating kennelled lying areas.

(i) Enclosed, insulated building with fan ventilation - of the 40% or so of herds in the sample which use this type of building it is reasonable to assume that a high proportion have partially slatted floors. Layouts do vary but a central feed passage with dunging channels along the outer walls is by far the most common. A wide selection of ventilation systems are used and it is only in the last few years that systems, incorporating accurate control of air-flow patterns which help to maintain cleanliness in such layouts, have come into general use.

(ii) Open-fronted mono-pitch - this is a system which was mainly straw-based until recent years. Now the front part of the pen is frequently slatted. Each pen is completely self-contained and all servicing takes place from outside. Ventilation control is by manual manipulation of front and rear ventilation flaps.

(iii) Kennel verandah - this term is used to cover all fattening houses which have a kennelled lying area and a slatted outside run. This is a development of the traditional cottage-type piggery. Ventilation control is by manual manipulation of the ventilation flaps.

(iv) Open, uninsulated building incorporating kennelled lying areas - in this system kennelled lying areas and 'runs' are covered by a clear span roof. The sides and ends of the building above kennel height are open or have space boarded cladding. Typically this system is used with straw bedding but a proportion have been built which incorporate slatted dunging areas. Ventilation control is by manual manipulation of the kennel lids.

3. FACTORS WHICH INFLUENCE THE OPERATION OF SYSTEMS INVOLVING PARTIALLY SLATTED FLOORS

The main factors are:

i. Climatic environment.
ii. Stocking density
iii. Pen design
iv. Proportion of slatted to solid floor area.
v. Slat specification

i. Climatic environment - The natural tendency of the pig is to keep the area on which it lies clean and dry and to dung and urinate elsewhere. However, it does have an overriding requirement to regulate heat loss in order to maintain a constant body temperature. As air temperature rises above comfort level the pig seeks to increase its heat loss and may do this by creating a wallow in the lying area or by lying in the dunging area.

The air temperature at which this occurs will depend to some extent on all the other factors which influence heat loss such as pig size, group size, feed level, air speed and the thermal transmitting properties of the floor. Livingstone and Robertson (1967) reported that air temperature adversely affected floor cleanliness at 21°C in pens containing bacon pigs. Heitman (1962) also found that where pigs had a choice their use of heat relieving measures increased rapidly when the air temperature was above 21°C.

In an investigation into the behaviour of fattening pigs Steiger (1979) found that where pigs were given the choice of floor surfaces their selection was related to air temperature. Taking account of the choices available within the investigation of straw bedding, solid floor and wooden slats it appeared that selection of the lying area was less determined by surface texture and softness than by climatic factors and the thermal conductivity of the floor.

The speed of air passing over the pig also influences its rate of heat loss. The pig reacts to variations in airspeed by avoiding or seeking out increased air speed depending on its current state of heat balance. Thus variable air speeds and/or air flow patterns have a considerable influence on the pig's choice of lying area. Ventilation systems which are unstable result in variable air speed at pig level which can cause the pig to keep changing its lying area in an effort to find a more comfortable place. Since the pig usually dungs away from its lying area a change in the lying site is often associated with a change in dunging site. Hence ventilation systems have a considerable influence on floor cleanliness in partially slatted systems. The high speed jet system advocated by Randall (1977) is particularly relevant to this problem.

ii. Stocking density - Livingstone and Robertson (1967) noted an increased tendency to dirty pens when stocking density exceeded or fell much below 0.55 m^2/baconer. Observation of commercial piggeries would support this conclusion. If the solid floored part of a partially slatted pen greatly exceeds the area required for lying there is every likelihood that dunging will occur on the solid area. A further conclusion of the study was that uninterrupted floor space, rather than total floor space, was an important factor in determining effective stocking density. Thus designs without physical barriers between solid and slatted areas were considered to allow relatively higher stocking density in relation to total pen size.

Where steps are taken to regulate solid floor space available to lying space requirements this is done in one of two ways:

a. Pig numbers are kept constant and the lying area is reduced by the use of an adjustable pen front.

b. Pig numbers are increased to fill the pen initially and reduced as growth progresses. The wide range of slaughter weights required in some parts of the United Kingdom can facilitate this policy.

iii. Pen Design - The Livingstone and Robertson (1967) study, which was only concerned with enclosed buildings, included observation of a comprehensive range of pen shapes and concluded that cleanliness was optimal where pen length to width ratio was 2 : 1 with the slatted area across the narrow axis. Slats across one or both ends of the pen produced clean solid areas but slats placed centrally resulted in the solid area at one end of the pen being soiled. A length to width ratio 3 : 1 is commonly used in North America and Randall (1975) observed that this ratio coincided with the cleanest pens. Observation of commercial practice indicates that length to width ratios of from 2 to 3 : 1 can be satisfactory in enclosed buildings.

Current commercial practice is to construct the slatted area approximately 50 mm below the level of the solid floor and to have a floor slope of about 1 : 40 towards the slats. A 25 - 50 mm step down to the slats is also reported by Randall (1975) as being widely used in North America. Slats placed slightly below the level of the solid floor assists pen cleaning and helps prevent the spread of liquid from the slatted area.

Watering facilities are almost invariably sited on the slatted area which undoubtedly assists in keeping the solid area clean. Watering devices cause wet floors which are unattractive as lying areas for the pig except at high air temperatures.

Feeding methods can have major influence on pen design. Current methods include *ad libitum* or controlled, wet or dry, trough, hopper or floor feeding. Troughs are sited either in the solid or the slatted area or in some cases, in enclosed buildings, troughs extend from one to the other. Controlled trough feeding imposes the most stringent constraints on pen shape in that trough space must be accurately related to lying area. Trough feeding is only feasible in some pen designs when feed is supplied by pipeline. Floor feeding and *ad libitum* hopper feeding allow more flexiblity of pen shape but care is needed to ensure that all pigs have easy access to the feed.

iv. Proportion of slatted to solid floor areas - current commercial practice is to provide a minimum width of slatted area of 1.07 m with a width of 1.2 m being quite common. Practical observations, supported by the Livingstone and Robertson study, suggest that slatted areas of less than 0.9 m wide give rise to soiling of the adjacent solid floor. United Kingdom welfare codes (1968) recommended a total space allocation of 122 kg/m^2 for pigs of over 57 kg liveweight. More recent draft recommendations suggest a lying area allowance of 0.45 m^2 for an 80 kg pig. If the total stocking density is taken as suggested above the total floor area for an 80 kg pig would be 0.65 m^2. If 0.45 m^2 is required as the lying area this leaves 0.2 m^2 available for dunging which is 30% of the total. Observation of commercial practice suggests that slatted areas down to 20% of the total can provide reasonably clean floors under good management but the task of maintaining cleanliness eases as the proportion of slatted area approaches 30%.

v. Slat specifications - a performance specification for animal flooring has been given by Baxter and Mitchell (1977) and also provides excellent guidelines for the requirements for slats:

i. Floors should not cause injury to livestock. They should provide a non-slip, non-abrasive surface, and should have no protruding edges.

ii. Floors should not contribute to ill health by harbouring bacteria and parasites. Surfaces should be impervious and readily cleaned.

iii. Floors should not contribute to 'stress' or discomfort which might manifest itself in depressed growth, poor feed conversion or abnormal behaviour. Floors should be matched thermally to the needs of the livestock housed on them.

iv. Floor materials and structures should not deteriorate or deform during their planned life, nor should they require maintenance during this period.

v. Perforated floors should not retain faeces or urine for periods of time which necessitate some form of cleaning or scraping.

vi. Floors should comply with the above requirements at the lowest acceptable cost.

The literature contains numerous reports of injuries to pig's feet caused by slatted and perforated floors of various types. Much of the damage has been caused by the use of inappropriate materials, poor design, manufacture, installation and maintenance. Although a variety of slatted and perforated materials are in commercial use the relative cost and durability of reinforced concrete slats makes them the most popular choice for fattening pigs. A slat width of 75 mm has been commonly used in the past but the more recent trend has been to use 100 mm slats in conjunction with 20 mm gaps. In a review of recommendations for concrete slats in Europe and the United States Baxter (1976) found that the majority came within the range of 75 - 125 mm slats and 18 - 25 mm gaps.

The Steiger (1979) investigation found that given the choice 30 - 90 kg pigs preferred wide (100 mm) wooden slats rather than narrow (50 mm) wooden slats for standing, walking and running. Observation of the pigs' walking ability confirmed that they walked substantially better on the wide slats. In a comparison of gap widths of 22 mm and 26 mm walking ability of 38 kg pigs appeared somewhat better with the narrower gaps. These observations are in good agreement with the recommendations attributed to Geyer (1978) of maximum gap widths of 18 - 22 mm for pigs at about 20 kg; 22 - 26 mm at about 40 kg and 24 - 28 mm at around 100 kg live weight.

The choice for fattening pens in commercial practice must depend on the weight range of pigs involved but under United

Kingdom conditions it would normally be prudent to assume that 20 kg pigs may occupy the pens at some stage. Thus the combination of 100 mm slats and 20 mm gaps would seem to be soundly based.

The quality of manufacture of slats is of crucial importance in order to ensure the absence of sharp or jagged edges. Pencil-rounded top edges in conjunction with non-slip, non-abrasive surfaces appear to be important.

The cleanliness of concrete slats is influenced by the width of slat, the width of gap and the slat profile. The proportion of slat to gap given above provides an acceptable level of cleanliness and appears to meet the requirements of the pigs. The side slope on the slat is of particular relevance to cleanliness and Baxter and Anderson (1976) recommend that this should be a minimum of one fifth.

4. CHARACTERISTICS OF PARTIALLY SLATTED SYSTEMS

A summary of the characteristics of the systems described is given in Appendix II. The intention is to illustrate that within the single category of partially slatted floors there is a variety of systems with differing characteristics which play a major part in influencing producers' choice of system and may also have some effect on the 'quality' of the environment provided for the pig.

The following characteristics apply particularly to enclosed systems:

i. A requirement for correct and stable air speeds and air flow patterns: inappropriate variations in air speed and air flow patterns cause pigs to change lying area which can lead to soiling of the solid area.

ii. A requirement for a carefully regulated air temperature: high temperatures - even within the thermoneutral zone - cause pigs to seek to wallow which leads to dirty floors.

iii. A requirement to keep the lying area correctly stocked: pigs choose their lying area and dung elsewhere. If a very low proportion of the lying area is actually occupied there is a greater likelihood of soiling on the solid area. Over-stocking also leads to dirty pens. Very high stocking rates are likely to impede the ability of the pig to regulate heat loss and may restrict access to the dunging area.

iv. A requirement to provide an appropriate pen shape: a length to width ratio of not less than 2 : 1 is indicated. Where there is no physical barrier between lying and dunging areas the slatted area should be across the narrow axis.

v. A requirement to provide an appropriate proportion of slatted area: a figure of 30% is likely to be optimum although designs involving only 20% can produce a reasonable level of cleanliness.

5. DO PARTIALLY SLATTED FLOORS ACHIEVE THEIR OBJECTIVE?

5.1. To minimise the workload

There is no doubt that well designed and managed systems do achieve a substantial reduction in labour input compared to solid floor systems, which utilise straw bedding. The percentage reduction depends on the total time which is influenced by all the component parts of the system. In a work study investigation of labour requirements of the housing systems used at the National Agricultural Centre, Meneer (1977) recorded a total time for feeding, cleaning out and littering down of open fronted mono-pitch pens of 1.75 min/pig/week. Of this total 0.8 min were required for cleaning and 0.33 min for littering down. The introduction of slats into this system eliminates the need for littering down and reduces cleaning time by an estimated 95%. The net effect would be a change of labour input for daily routine tasks from 1.75 to 0.66 min/pig/week which is a reduction of 62%.

Using a combination of recorded data and extrapolation it would seem likely that in a kennelled housing situation the introduction of slats in place of solid floors and bedding would reduce labour requirements from 1.55 to 0.64 min/pig/week which is a 59% reduction. Thus the time saving likely to be achieved in the two systems described is of the order of 1 min/pig/week. Another factor which needs to be considered is the reduction in the menial element which contributes to the quality of the job.

5.2. To eliminate the cost of straw bedding

The quantity of straw used varies by system, season and stockman and price varies by geographical location and time. It has been calculated by Brent (1975) that straw usage accounted for between 4 and 5% of production costs at the National Agricultural Centre. These figures refer to open fronted mono-pitch and to kennelled housing and are based on straw requirements of 0.50 and 0.54 kg of straw/kg of liveweight gain. In a subsequent year straw usage was 0.63 kg and 0.81 kg/kg liveweight gain. In each case the higher figure refers to the kennelled housing. Taking a mean figure for straw use of 0.65 kg/kg liveweight gain, a liveweight increase of 54 kg and a straw price of 3 p/kg gives a straw cost of £1.05 per finished pig. In practice such costs are sometimes set against the fertiliser value of the straw manure but this value is not always reliable.

In situations where it is possible to introduce partially slatted floors in place of solid floors without involving changes of dimension a slight increase in capital costs would normally be involved. Currently open fronted mono-pitch pens with partially slatted floors cost 21% more than pens with solid floors (£51 compared to £42/pig place). This represents an overhead cost of 45 p/pig allowing for 4 batches per year over a 10 year life at 15% interest rate.

It must be clearly understood however, that solid floor housing systems are available across the entire cost range from £42 to over £100/pig place and therefore many solid floor systems involve higher capital costs than partially slatted systems.

A considerable weight of scientific evidence produced by Mount (1967), Verstegen (1974) and others demonstrates that the thermal conductivity properties of floors have a considerable influence on the air temperature requirements of pigs. Bruce (1979) has developed a method of measuring and classifying the thermal characteristics of floors and subsequently incorporated these data in mathematical models (Bruce and Clark, 1979) which can be used to determine critical temperatures for growing pigs. Figures from the model show a lower critical air temperature increase of approximately 6°C where straw bedding is substituted by a solid concrete floor and an increase of 1½ - 2°C where straw is substituted by an insulated, concrete floor. This emphasises the fact that the absence of straw bedding demands increased attention to climatic control particularly during cold ambient conditions.

Published comparative data fail to show any consistent differences between pig fattening systems which cannot be explained by nutritional, health or environmental factors. In a review of research on the influence of floor design on pig performance, Jensen (1978) concluded that the growth rate and feed conversion efficiency in growing/finishing pigs has in general been comparable on partially and totally slatted floors. Where sizeable differences have occurred factors other than the slatted floor were likely to be primary. Effective air temperature, stocking density, group size, design, construction or installation of slats were among such factors.

There is a widely held subjective view that partially slatted floors may have a favourable influence on pig health. If such benefits do occur it would seem that they are most likely to do so where the comparison is with sweep through solid systems where newly weaned or purchased pigs share a common dunging passage.

A positive benefit of systems which incorporate properly designed and constructed partially slatted floors is the element of choice given to the pigs. By their behaviour stockmen can readily judge whether the environment meets their needs and take steps to make appropriate changes to keep the lying areas clean.

5.3. Partially slatted floors used in conjunction with straw bedding

This combination appears to have a number of theoretical attractions. The use of straw would alleviate the effects of a marginally poor climatic environment and also probably enrich the overall environment for the pig. In the Steiger (1979) study pigs were given a choice of flooring and their well known preference for straw covered floors was confirmed. In commercial practice small quantities of straw are occasionally used, in partially slatted pens, to counteract outbreaks of tail biting rather than as an integral part of a system. There have been a few reports (Farm Buildings Progress, 1971) of purpose built straw/slat systems but they are not in common use in the United Kingdom. The reasons for the lack of enthusiasm for this system would include the cost and availability of straw, the extra labour required for littering down and cleaning out, the fear of slats blocking up and the possible difficulty of removing relatively dry manure from the slurry pit. Good design would, no doubt, minimise the latter problems but the lack of straw in some areas and the desire to minimise menial tasks would remain as a deterrent to many.

APPENDIX 1

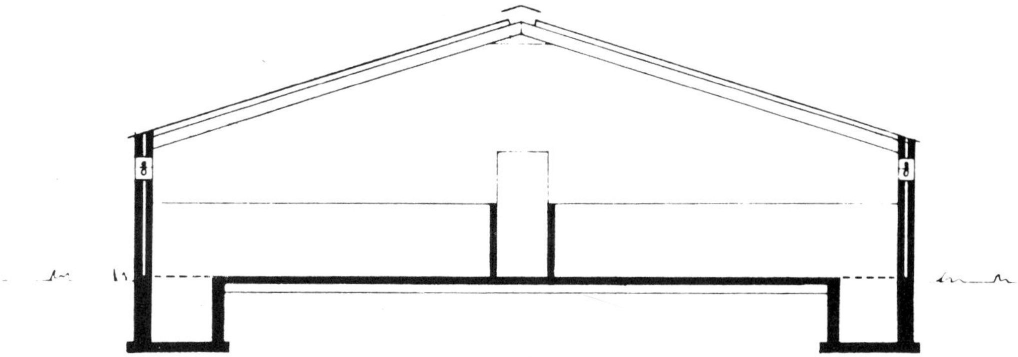

Enclosed, insulated, with controlled ventilation.

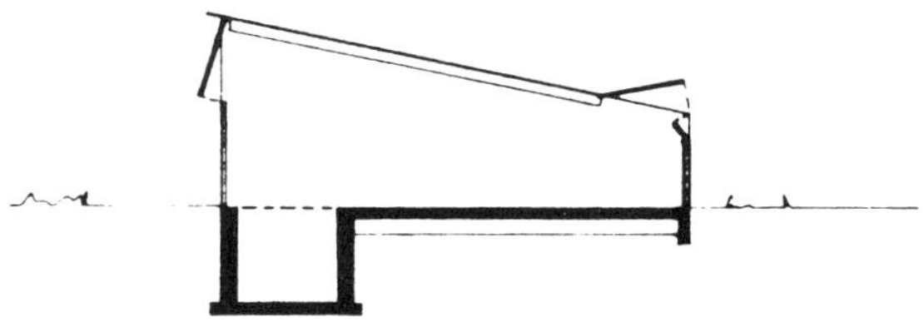

Open fronted mono-pitch.

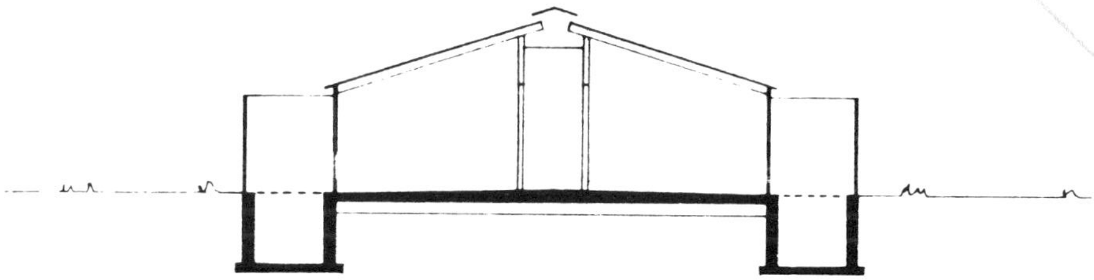

Kennel verandah.

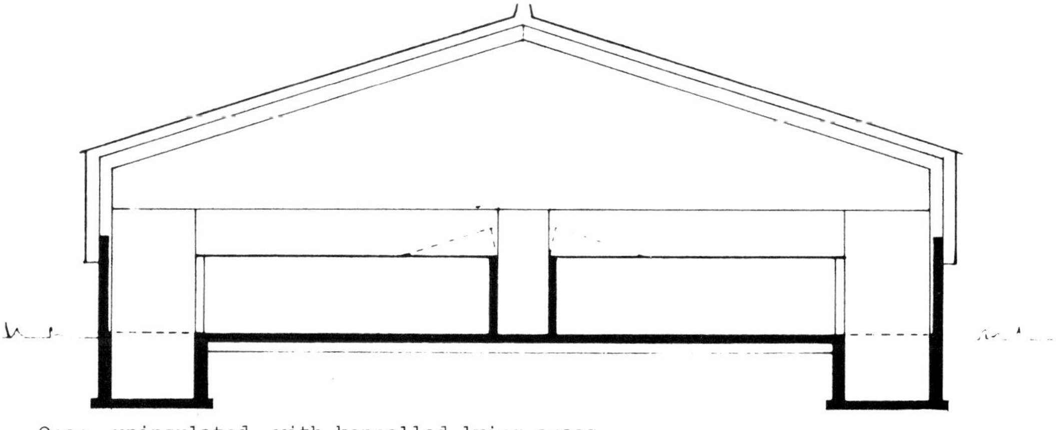

Open, uninsulated, with kennelled lying areas.

APPENDIX 2

A COMPARATIVE ASSESSMENT OF THE CHARACTERISTICS OF PIG FATTENING SYSTEMS INVOLVING PARTIALLY SLATTED FLOORS

System	Capital cost 1	Labour 2	Working conditions 3	Choice of environment for the pig 4	Temperature control 5	Reliance on mechanical controls 6	Dependency on stockmanship 7
Fully enclosed, insulated with automatically controlled ventilation	1.0	++	+++	+	+++	+	+++
Open, uninsulated building incorporating kennelled lying areas	1.23	+	++	+++	+	+++	+
Kennel verandah	0.83	+	++	+++	+	+++	+
Open fronted mono-pitch	0.83	+	+	++	+	+++	++

The more +s the better. Comparison is with fully enclosed systems

1. Initial cost.
2. Total labour input.
3. Assessment of working conditions for the stockman
4. Choice of environment available to the pig under good conditions.
5. Probable degree of temperature control under good conditions.
6. Reliance on mechanical controls (∴ need for fail safe arrangements).
7. Extent to which the system depends on better than average stockmanship.

The data in this Table are merely indicative of trends

REFERENCES

Baxter, S.H. and Anderson, A.W.F., 1976. A new reinforced slat system for pig housing. Farm Building Progress, 46.

Baxter, S.H. and Mitchel, C.D., 1977. Developments in floor construction in animal production. Veterinary Annual, 17, 286-291.

Brent, G., 1975. National Agricultural Centre, Pig Unit Report.

Bruce, J.M., 1979. Heat loss from animals to floors. Farm Building Progress, 55.

Bruce, J.M. and Clark, , 1979. Models of heat production and critical temperature for growing pigs. Animal Production, 28.3.

Codes of Recommendations for the Welfare of Livestock Code No 2 Pigs, 1968.

Farm Building Progress 26, 1971. Straw-bedded weaner pool with slats.

Geyer, H., 1978. Morphology and Growth of Pigs feed - thesis. Faculty of Veterinary Medicine, Zürich.

Heitman, H., Hahn, L., Bond, T.E. and Kelly, C.F., 1962. The effects of a modified summer environment on swine behaviour. Animal Behaviour 10, 15-19.

Jensen, A.H., 1978. The effects of environmental factors, floor design and materials on performance and on foot and limb disorders in growing and adult pigs. Symposium on injuries due to floors in animal housing.

Livingstone, H.R. and Robertson, A.M., 1967. Partially slatted floors and floor feeding in Pig Fattening Houses. Experimental Farm Buildings Report No 8.

Meat and Livestock Commission Hit Count, March 1979.

Meneer, R., 1977. Unpublished report.

Mount, L.E., 1967. The heat loss from new born pigs to the floor. Res. Vet, Sci., 8.

Randall, J.M., 1977. A handbook on the design of a ventilation system for livestock buildings using step control and automatic vents. NIAE Report No. 28.

Randall, J.M., 1975. Environment and Welfare aspects of Intensive Livestock Housing. A report of a North American tour.

Stieger, A., Tschanz, B., Jacob, P. and Scholl, E., 1979. Investigation of behaviour of fattening pigs on different types of floor and with different densities. Schweiz. Arch. Tierheilk.

Verstegen, N.W.A. and Van der Hel, W., 1974. The effects of temperature and type of floor on metabolic rate and effective critical temperature in groups of growing pigs. Animal Production 18.1.

DISCUSSION

G. van Putten *(Netherlands)*

I am very interested in the number of pigs with leg weakness, for instance, in the systems that you compared. Have you any information concerning that characteristic?

A.T. Smith *(UK)*

No, I am afraid that I have no statistical data on that at all. I would observe that such differences that one found within these systems would be a reflection of the type of floor, type of slat, the quality of manufacture and installation

J. Daelemans *(Belgium)*

You stated that pigs choose their lying area and dung elsewhere. That is so, but there is a difficulty in summer when the temperatures are high. Normally results are better with partially slatted pens, but during summer our pigs are turning and mucking on the wrong side and lying on the slats for coolness. Have you any solutions for this, because the pigs get very dirty in summer which is not desirable?

A.T. Smith

You may have much greater experience of high temperatures, it is not a problem that we have to the same extent as you do. However, my impression is, at least in the UK, that we have not as yet made full use of the direction in which we aim our ventilation systems. When we established some of these high speed ventilation systems, we did envisage the need to change the direction for the air flow in order to relieve heat stress.

D.L. Ingram *(UK)*

Not being a practical farmer, I would like to ask whether a dirty pig is less commercially viable than a clean one, or is this an aesthetic matter? You asked whether a pig should be deprived from the opportunity to wallow, is this frustrating natural behaviour? I think that the answer must be that if it is hot and you stop him from wallowing, you are then frustrating him. If you allow him to wallow, you said that it displeases the pig man, but is it commercially less viable?

A.T. Smith

There are a whole range of answers to this, depending on the period of time involved, and just how dirty the pig becomes.

When for instance wet and muddy pigs experience a sudden air temperature drop the critical temperature level might be surpassed.

OPEN FRONT PIGGERIES WITH AND WITHOUT STRAW

P. Koomans

Institute of Agricultural Engineering, PO Box 43,
6700 AA, Wageningen, the Netherlands

1. INTRODUCTION

In the last few years there has been an accent on saving on building and labour costs in the development of housing systems for fattening pigs. The areas for each animal are getting smaller and smaller. In general there is no straw available any more. In some cases the limits of adaptability of the animals have been exceeded. The results are symptoms of stress, illness and finally mortality. One of the signs that adaptability has broken down is the occurrence of cannibalism.

On the other hand it is well known that straw in pens has a favourable influence on the well-being of pigs.

By keeping the animals in an open front piggery one can economise on heating and on electricity for ventilation.

To obtain more information about both of these aspects and their effect on production, different experiments have been set up.

2. MATERIALS AND METHODS

An open front piggery with straw and an open front piggery with a partly slatted floor were used for the experiments. Each piggery had 4 pens for 15 animals each. In both piggeries the animals were fed *ad libitum*. The area/animal was 1 m^2 including the feed hopper. In the piggery with a partly slatted floor the area/head was 0.6 m^2 solid floor and 0.4 m^2 slats. In this piggery a part of the solid floor had a ceiling of 3 m^2 made of insulating board at 1.20 m above floor level. For the animals a protected lying area under the board was provided.

Water was available for the animals all day long. Once or twice a week straw was replenished in the pens. Dung was removed at the end of the fattening period. The pens were also cleaned and disinfected. The open side had a gap 1.5 m high over the total width of the pen. Only during strong frost was a plastic curtain suspended from the ceiling in the middle of the pens of the piggery with partly slatted floor. The drinking nipples were protected against frost by an infra-red heating bulb. The most important figures are given in Table 1.

TABLE 1

AVERAGE FEED CONVERSION, DAILY GAIN AND BACKFAT OF THE PIGS FATTENED IN AN OPEN FRONT PIGGERY WITH STRAW AND IN AN OPEN FRONT PIGGERY WITH A PARTLY SLATTED FLOOR.

	Open front piggery with straw	Open front piggery with partly slatted floor	
Number of animals	300	300	
Average feed conversion	3.35	3.37	n.s.
Average daily gain in g	811	742	xxx
Average backfat in mm	36.2	34.4	x
Amount of straw/animal/ fattening period	80-100	-	
Animals culled	5	11	

The figures are corrected for a startweight of 23 kg and a slaughtered weight of 83 kg. With the exception of the average backfat thickness the results in the open front piggery with straw were better than those in the piggery with a partly slatted floor.

The number of animals culled in the piggery with straw was half as much as in the piggery with partly slatted floor. In the latter piggery the reason for the culling of 8 animals was cannibalism. In the piggery with straw no cannibalism was found. Two pigs in the piggery with straw and 5 in the piggery without straw were found to be lame. The open front piggery without straw was found to be the more economical of the two, mainly because of the costs of straw and the costs of muck removal.

3. STRAW EFFECT

For this reason a variation was introduced into the experiment. It is better for young pigs to be kept in straw. When the pigs reached a weight of about 60 kg, they were moved to a pen with a partly slatted floor without straw. The effect of moving the animals from the pen with straw to a pen without straw was measured in the following experiment.

The pigs were divided into 3 groups, each consisting of animals with the same startweight, sex and genetic background. One group was kept in a pen with straw throughout the experiment.

Another group was fattened from start to finish in a pen with a partly slatted floor; the third group of animals started in a pen with straw and were moved at 60 kg to a pen without straw.

The effects of moving the animals might be improved carcass quality, a saving on litter and a saving on expenses for dung removal. The disadvantages were the extra labour for cleaning and disinfection, a reduction in the well-being of the animals and a somewhat unfavourable figure for daily gain. In Table 2 the most important figures are given.

The figures were also corrected for a startweight of 23 kg and a slaughtered weight of 83 kg.

In comparison with the first experiment the culling rate was reduced. In the open front piggery with partly slatted floor and in the open front piggery into which the pigs were moved, respectively 3 and 1 animals died from cannibalism. In the open front piggery with a partly slatted floor 3 animals had a slower rate of growth.

TABLE 2

AVERAGE FEED CONVERSION, DAILY GAIN AND BACKFAT OF PIGS FATTENED IN AN OPEN FRONT PIGGERY (A) WITH STRAW, (B) PARTLY SLATTED FLOOR AND (C) STARTED IN A PEN OF SYSTEM A WITH STRAW AND MOVED AT 60 kg LIVE WEIGHT TO A PEN OF SYSTEM B.

	Open front piggery with straw (A)	Open front piggery with partly slatted floor (B)	Open front piggery with straw. Pigs moved at 60 kg (C)	Differences		
Number of animals	240	240	240	A-B	A-C	B-C
Average feed conversion	3.09	3.11	3.09	n.s	n.s	n.s
Average daily gain	846	789	804	xxx	xx	x
Average backfat in mm	32.0	30.3	30.6	xxx	xx	x
Amount of straw/pig/ fattening period	80-100	-	30-40			
Animals culled	2	6	2			

The figures in Table 2 show the same tendency as those in Table 1. The results for the pigs in the group which was moved were between those for the piggery with straw and those for the piggery with partly slatted floor.

4. ECONOMIC DATA

The differences in feed conversion, daily gain, backfat, housing and labour are given in financial terms in Table 3.

TABLE 3

FINANCIAL DIFFERENCE IN GUILDERS/SLAUGHTERED PIG FATTENED IN THE OPEN FRONT PIGGERY WITH STRAW WITH RESPECT TO THE OPEN FRONT PIGGERY WITH PARTLY SLATTED FLOOR AND THE OPEN FRONT PIGGERY WITH STRAW, INTO WHICH THE ANIMALS ARE MOVED AT 60 kg.

	Open front piggery with partly slatted floor		Open front piggery with straw. Pigs moved at 60 kg.	
	Positive	negative	positive	negative
Housing		0.75		0.38
Daily gain		2.38		1.75
Feed conversion		0.95		
Carcass quality	1.70		1.40	
Straw	12.00		4.50	
Dung removal	3.00		1.50	
Cleaning + disinfection				1.00
Culling		3.32		
positive	16.70		7.40	
negative		7.40		3.13
total	9.30		4.27	

The conclusion to be drawn from Table 3 is that from the point of view of well-being the open front piggery with straw is attractive, but it is less economical. There is still a need for the development of new housing systems in which well-being and economy coincide.

5. CONCLUSIONS

Two experiments are described in which pigs are fattened in an open front piggery with straw and in an open front piggery with partly slatted floor.

The following conclusions can be drawn:

I. The piggery with straw is more favourable than the piggery without straw from the point of view of the well-being of the pigs.

II. Disadvantages of the piggery with straw are the extra costs of straw and of labour for dung removal.

III. The yield/pig in the open front pighouse with straw was approximately D.fl.9, - less than in the piggery with partly slatted floor, in spite of the fact that in the latter one more animal died than in the former.

DISCUSSION

I. Ekesbo *(Sweden)*

It is interesting to try to translate results into guilders, pounds, etc., but against the background of Table 3 and its conclusions, what are your explanations for the findings in Tables 1 and 2 concerning backfat in mm. In relation to the open front piggery with straw, in Table 1 there is 36.2 mm of backfat, and in Table 2 there is 32.0 mm of backfat. In relation to the open front piggery with partly slatted floor, in Table 1 there is 34.4 mm of backfat, and in Table 2 for the same group there is 30.3 mm of backfat. How sure are you that the differences found in these two pairs of figures are real differences caused by the two systems?

P. Koomans *(Netherlands)*

Firstly, the difference between Tables 1 and 2 is that the experiments began in 1975 and ended a few months ago; also selection took place in the sow herd on the experimental farm so that in general the quality of the sows was higher. Secondly, I think that these figures are comparable because in both experiments the pigs were divided into groups using such criteria as sex, heredity and weight. I think the conclusions are valid that the differences are due to the systems.

A.T. Smith *(UK)*

What climatic analyses were carried out? Do you know what the temperature and air speeds were over the pigs in both pens?

P. Koomans

The air velocity was the same in both pens. The first slide showed that there was really little difference between the two pens. In summer the temperatures were of a normal high level, and in winter the temperature in the pens was $-6^{o}C$ to $-7^{o}C$.

A.T. Smith

Were the differences between the two systems consistent in summer and winter?

P. Koomans

Yes. In the first experiment there were 60 animals in one group. In Table 1 there were 5 repeats, and there were 3 groups in winter and 2 in summer. In Table 2 there were 30 animals, with 8 repeats over a 2½ year period. In Table 2 there were more repeats in summer than in winter.

D.L. Ingram *(UK)*

I think that this paper makes a point that this seminar ought to underline - that welfare is likely to cost money. It is no use saying that we must do more research to find an ideal solution, it may not be there, and so it is up to people to decide whether or not they are prepared to pay the cost of animal welfare. If we fail to do that, we will be failing in our purposes.

Doris Buchenauer *(FRG)*

In your calculations, the cost of building the stables is not increased. If you calculate the cost of the stable by what, in Germany, is called the price per animal, what would be the result?

P. Koomans

There is a difference in the cost of housing of about 15% in investment per pig place between a pen with a partly slatted floor and a pen with straw.

Doris Buchenauer

Therefore, when carefully calculated, partly slatted floors are more expensive than solid floors?

P. Koomans

Yes.

I. Ekesbo

If the conclusions are right, which might be possible bearing in mind the additional information you have given in relation to the backfat question, how do you explain the difference in backfat in these two systems? This is an interesting biological question.

P. Koomans

Feed intake is important. The feed conversion rate in both systems was the same; there was a big difference in daily gain so the average feed intake per day makes a difference of perhaps 200 - 300 g/day, maybe a little more. This may account to a large extent for the difference in backfat.

I. Ekesbo

There is no significant difference in the feed conversion capacity, as far as I can understand.

P. Koomans

Yes, that is true, but I think that to a large extent you can link the difference in backfat to the difference in intake.

G. van Putten *(Netherlands)*

In Table 3 you compared two housing systems: the open front piggery with the partially slatted floor, and the open front piggery with straw from which the pigs were removed. The removal of the pigs must be quite a job and presumably entails having an additional housing system. Can you tell me what the costs would be if the pigs were not removed?

Secondly, I do not think that the amount of backfat is the important question; the important factor is the amount of valuable meat that remains, excluding fat.

P. Koomans

In answer to your first question, if you leave the animals in straw then you find that the difference in total cost is D. fl. 9.30 per animal.

In reply to your second question about the meat quantity, I reckon that there will not only be financial differences but you have to use daily gain and feed conversion. The only correction factor I have for backfat is one Dutch florin (guilder) per 0.3 mm differences. I have no figures about the amount of meat.

J.P. Signoret *(France)*

To return to Dr. Ingram's comment that we have to pay for animal welfare: we must be careful not to make the assumption that welfare means straw, and that welfare means greater expense. Occasionally, as stated by Dr. Unshelm, welfare and economic returns can well go together, in other cases they evolve in opposite ways. Sometimes we have to pay for welfare and in other cases welfare will pay and give higher returns.

P. Koomans

So we must try to find new housing systems in which the interests both of welfare and economy are served.

K. Vestergaard *(Denmark)*

If you look at the count there is about 57 g more daily gain in the open front piggery with straw as compared with the open front piggery with partly slatted floor. This means that the animals are within the system for a shorter period and you can use the house more efficiently. Have you taken this into consideration in regard to the economics of the system?

P. Koomans

Yes.

J.M. Bienfait *(Belgium)*

You use 80 - 100 kg of straw per pig. Are you sure that the results are better with this quantity? In Belgium we use 30 - 50 kg per pig.

P. Koomans

It is not practical to use less straw. The pens become dirty if you use less than 80 - 100 kg of straw, and this would not be acceptable to the farmer. Generally, more straw is needed in summer than in winter.

SESSION IV

COMPARATIVE DATA REGARDING WELFARE IN PRESENT HUSBANDRY SYSTEMS, I.

2. PIGLETS

Chairman: R. Moss

WELFARE OF PIGLETS IN FLATDECK-CAGES

G. van Putten
Institute for Animal Husbandry "Schoonoord",
Zeist, The Netherlands

1. INTRODUCTION

According to Lorz (1973), welfare means: 'living in harmony with the environment and within the animal itself'. This 'well-being' can be expressed in terms of health and of behaviour. Nowadays physiology and a method developed by the Swedish veterinarian, Ekesbo, as well as production, are being added to the parameters. The latter, I admit, is being added with some reluctance. We are even able to express welfare in figures. However, there is one restriction: these figures are not absolute, but only relative. This means, that if we speak of welfare as being influenced by a certain type of housing and husbandry, we can only do so in relation to another system. This handicap, if you want to call it so, also applies to piglets kept in cages.

2. AIMS

The aims of this paper are threefold: firstly, to look into the future and to predict whether or not early weaning will be continued in the years ahead; secondly, to ask ourselves whether it is really possible to reject flatdeck-cages; and thirdly, which alternative housing systems to flatdeck-cages might better ensure the welfare of piglets.

3. EARLY WEANING

Very early weaning of piglets has been tried. Nearly every age between 3 days and 60 days after birth has been tested on its economic merits. This was the era of battery-cages and a golden future was predicted for those who used them. In reality they were not that profitable because the sows did not co-operate.

If piglets were weaned under the age of 3 weeks, the fertility of the sows involved declined so that all efforts to make more money were in fact in vain: on the contrary, weaning before the age of 3 weeks obviously costs money.

Lively discussions about the best time of weaning between 3 and 6 weeks still continue. We shall come to that later. One thing is very clear already: weaning at the age of 8 weeks, as farmers used to do some 20 years ago, costs too much money to become common practice once more. Thus some kind of housing and husbandry for early weaned piglets will have to be accepted, whether we like it or not.

4. ARE FLATDECK-CAGES REALLY REJECTABLE?

To answer the question of whether it is really possible to reject flatdeck-cages on the grounds of the welfare of piglets kept in these cages, we have to compare them with other practicable systems, as has already been explained in the Introduction.

The following authors compared the well-being of piglets kept in flatdeck-cages with that of piglets kept in other systems: Algers, 1979; Danielsen and Nielsen, 1978; Fiedler, 1973; Fraser, 1978; Loonstra, 1973; van Putten and Dammers, 1976; Schnieder and Bronsch, 1974. All of them expressed the conviction that some other housing system ensures the welfare of early weaned piglets better than flatdecks do. We shall analyse their data later in this paper.

Besides welfare there are other points to be taken into consideration in judging flatdeck-cages. Both Bogner and Puff (1978) and Marx and Schuster (1980) state that it is very difficult to design a good cage, and to maintain a good climate within that cage. The type of floor is crucial to cage design. Marx and Schuster (1980) allowed the piglets to indicate the floor of their choice from among 6 types. They preferred equally a slatted concrete floor, a perforated metal floor and a special type of plastic floor, whereas they obviously avoided the wire-mesh floor. Bogner and Puff (1978) also found that the combination of the type of cage with the right technical outfit is very important. They stress that the climatic conditions in a room with cages have to be kept within very narrow limits. The conclusion can only be that cage-rearing of piglets is a very vulnerable system and thus asks for constant care.

Flatdeck-cage systems for rearing piglets may be difficult to design, but there is yet another problem. Two authors (Otto and Kalm, 1978; Bogner and Puff, 1978) report that piglets originating from flatdeck-cages do not have as good a performance as those kept with the sow for 7 weeks. Otto and Kalm (1978) compared 390 pigs weaned at 21 days, with 331 pigs weaned at 42 days, and found a significantly higher slaughterweight of 0.9 kg in the late weaned group.

5. WHAT ARE THE DISADVANTAGES OF FLATDECK-CAGES REGARDING WELFARE?

Apart from the vulnerability of the system and the poorer performance of piglets in flatdeck-cages, the welfare of the animals involved is causing problems. Which components in the flatdeck-housing system actually cause this lack of well-being in the piglets may become clear when comparing available results of experiments.

Van Putten and Dammers (1976) compared the welfare of 8 litters, weaned at an age of 3 - 4 weeks and kept on flatdeck-cages, with that of 8 litters kept with the sow on straw.

Every week in each housing system, around the clock observations were made with records of 16 behavioural patterns every 2 min for the individual piglets. These showed a significantly higher incidence of abnormal behaviour in the litters in a poor environment on a flatdeck-cage, than in the litters in a rich environment with the sow in a pen with straw.

Even after having been on the flatdeck for 3 weeks it was shown that the piglets missed the stimuli given by the sow, and this resulted in conflict behaviour, expressed as redirections from massaging the udder and sucking the teats of the sow towards other piglets and lifeless objects. Significantly higher incidences of exploratory behaviour in the pen and redirection towards other objects in the cages demonstrated a lack of adequate stimuli in the flatdeck system and particularly a lack of straw. Fraser (1978) confirmed the conclusions of van Putten and Dammers (1976) in comparing 4 unweaned litters with 4 weaned litters on wood-shavings.

Though van Putten and Dammers (1976) came to a distinct conclusion about welfare, they were unable to discriminate clearly between the contribution of the sow and the contribution of straw. Algers (1979) compared the behaviour of 162 piglets, weaned at 3 weeks and kept in flatdeck-cages, with that of 88 piglets also weaned at 3 weeks but kept on straw.

He found markedly greater tendencies towards massaging other piglets, sucking other piglets and nibbling on other piglets in the flatdeck-cages. As he found similar patterns of behaviour in the flatdeck-cages as van Putten and Dammers (1976) did the differences between the pens of Algers and those of van Putten and Dammers must be due to the absence of the sow. This means that the remaining differences between the piglets in the pens and those in the cages of both Algers (1979) and van Putten and Dammers (1976), such as increased sitting like a dog, less resting, more massaging of lifeless objects and more nibbling on lifeless objects in the cages, were caused by the absence of straw.

Quite a number of experiments were carried out in order to find an acceptable substitute for straw in the pens for early weaned piglets. Schneider and Bronsch (1974) weaned at an age of 3 weeks and kept 169 piglets in cages, 163 piglets in pens with insulated concrete floors covered with synthetic materials and 166 piglets in pens with insulated concrete floors covered with straw. During the first 3 weeks after weaning, the piglets in the pens did equally as well as, or better, than those in the cages. Their food conversion was 106% compared with the cages. In the flatdeck-cages morbidity rate (diarrhoea) was higher: 2%. The piglets on synthetic materials had a morbidity rate of 1.7% and those on straw 1.5%. Mortality was highest on synthetic materials: 3%. In the cages 1.1% of the piglets died and on straw 1.8%. The conclusion to be drawn from this experiment is that a concrete floor covered with synthetic materials cannot compete with a concrete floor covered with straw.

Danielsen and Nielsen (1978) compared 169 piglets in flatdeck-cages with 138 piglets on a concrete floor, covered with shavings. The piglets were weaned at 3 weeks. If we equalise morbidity and mortality of piglets on flatdecks found by Danielsen and Nielsen (1978) with the percentages found by Schneider and Bronsch (1974), we find a morbidity rate of 1.6% and mortality rate of 1.3% in piglets kept on shavings. This means that shavings and straw show minor differences regarding the welfare of piglets kept on this bedding.

6. WEANING AGE

One of the difficulties in judging flatdeck-cages in relation to welfare is the difference in weaning age. Te Brake (1978) advocated weaning at an age of 21 - 25 days as being the most profitable. Loonstra et al. (1973) did not find any difference in financial profit between weaning at 3 and weaning at 5 weeks. Oudenampsen (1974) is very much in favour of weaning at 5½ weeks instead of 3½ weeks, because the former has a better food conversion in the rearing pens. Danielsen and Nielsen (1978) found a distinct correlation between the weaning age and the rates of mortality and morbidity (Table 1).

The choice of weaning-time is extremely difficult. In practical farming one can observe a tendency to wean at the age of 5 weeks. This situation, combined with the established fact that welfare is less secure in flatdeck-cages than in most other systems, makes one ask whether the maintenance of flatdecks for this relatively short period between weaning and fattening is still worthwhile. This question can only be answered by considering the alternatives.

TABLE 1

RELATIONSHIP BETWEEN WEANING AGE AND MORBIDITY AND MORTALITY OF PIGLETS KEPT IN FLATDECK-CAGES WITH A FLOOR OF EXPANDED METAL. (DANIELSEN AND NIELSEN, 1978).

Weaning age (weeks)	Number of litters	Mortality, in %	Morbidity (diarrhoea), in %
3	20	0.6	1.3
4	19	0.9	1.3
5	22	1.0	1.0
6	81	0.0	1.1

7. ALTERNATIVE HOUSING SYSTEMS

Alternative housing systems for weaned piglets should meet 3 demands:

- ensure a better welfare of the animals involved than flatdeck-cages;
- be financially competitive with flatdeck-cages;
- be practicable.

Again 3 housing systems deserve consideration. First, there is the pen with an insulated concrete floor covered with straw. Algers (1979) pleaded in favour of a pen of this type and so did Schneider and Bronsch (1974). Oudenampsen (1974) found a slightly better food conversion in 376 piglets reared in pens with straw than in 667 piglets kept in flatdeck-cages.

Secondly, there is the very attractive possibility of leaving the weaned piglets in the (enlarged) farrowing-rearing pen enriched with straw on the insulated concrete floor, after having removed the sow. In 1973, a year of very low energy prices, Loonstra et al. had already calculated that weaning the dam at 3 weeks and rearing the piglets in the farrowing pen did not cost more than placing them on flatdeck-cages at the same age. Van Ingen and Koomans (1980) compared 7 types of farrowing-rearing pens. Their piglets were weaned at an age of 3½ weeks. Their conclusion was that a fully or partially slatted concrete floor was not suitable for a farrowing-rearing pen, that a conrete floor with straw obviously had the lowest morbidity in combination with a good food conversion, and that a number of metal floors still had to be better evaluated. In the opinion of these authors a farrowing-rearing pen is too expensive because the piglets have much more floor-space after weaning than is needed.

The third alternative is an open front - deep litter rearing pen, designed by Koomans. Troxler (1979) presented some data on a comparison between this type of rearing pen and the traditional flatdeck-cage. The piglets were weaned at an age of 4 weeks. One hundred and seventy piglets in the open front - deep litter pen were compared with 167 piglets in a flatdeck-cage. In the open front pen both growth and food conversion were better and morbidity was lower. Ethological 24 h observations also indicated that welfare in the open front - deep litter pen was much better.

The obvious conclusion to be drawn from this is that early weaning does not inevitably imply that the piglets should be kept in flatdeck-cages. There are alternative systems available which can compete economically with flatdeck-cages and have less disadvantages regarding the welfare of piglets.

8. SUMMARY AND CONCLUSIONS

I. There seems to be no reason to regard early weaning of piglets as a passing fashion. However, a tendency can be observed to postpone weaning until the piglets are about 5 weeks old. This means that only 3 or 4 weeks are left until the piglets are ready to be placed into fattening-pens.

II. Intensive experiments in a number of countries made it clear that the welfare of piglets kept in flatdeck-cages is at a relatively low level when compared with the welfare of piglets kept in other housing systems. The differences are expressed in terms of behaviour, of health and of production.

III. Practical alternative housing systems are available with which to replace flatdeck-cages and bridge the gap of 3 or 4 weeks between weaning and fattening.

REFERENCES

Algers, B., 1979. Gesundheit und Verhalten von Ferkeln in Flat-Deck-Haltung. Vortrag gehalten an der Informationstagung "Tiergerechte Nutztierhaltung" für die Internationale Gesellschaft für Nutztierhaltung (IGN), 31 October, Basle, Switzerland.

Bogner, H. and Puff, H., 1978. Das Frühabsetzen von Ferkeln. Bayer. Landw. Jahrbuch, 55 (5): 522-530.

Danielsen, V. and Nielsen, H.E., 1978. Rearing of early weaned piglets on flatdecks or on concrete floors. Paper 29th Annual Meeting of the EAAP, Commission on Animal Management and Health/Pig Production, Stockholm.

Fiedler, E., 1973. Flats für Ferkel. Top Agrar., 5: 1-5.

Fraser, D., 1978. Observations on the behavioural development of suckling and early-weaned piglets during the first six weeks after birth. Anim. Behav., 26: 22-30.

van Ingen, T. and Koomans, P., 1980. Vergelijking van kraamopfokhokken. I. Sterksel: Varkensproefbedrijf Zuid- en West-Nederland. Proefverslag, No. 17.

Loonstra, A.J., Kemkers, G. and Vreeman, H.J., 1973. Spenen van biggen op 3, 5 of 7 weken. Wormerveer: Wessanen. Rapport Landbouwk. Afd. No. 4.

Marx, D. and Schuster, A., 1980. Ethologische Wahlversuche mit frühabgesetzten Ferkeln während der Flatdeckhaltung. Dtsch. tierärztl. Wschr., 87: 365-400.

Otto, H. and Kalm, E., 1978. Der Einfluss der Aufzuchtmethode von Ferkeln auf deren spätere Mast- und Schlachtleistungen bei unterschiedlicher Haltung während der Mast. Paper 29th Annual Meeting of the EAAP, Commission on Pig Production, Stockholm.

Oudenampsen, H., 1974. De opfok- en mesterijresultaten van op 3½ week gespeende en naar grondhok of batterij verplaatste biggen, en van op 5½ week gespeende en niet verplaatste biggen. Sterksel: Varkensproefbedrijf Zuid- en West-Nederland. Proefverslag, No. 3.

van Putten, G. and Dammers, J., 1976. A comparative study of the well-being of piglets reared conventionally and in cages. Appl. Anim. Ethol., 2: 339-356.

Schneider, D. and Bronsch, K., 1974. Aufzuchtvergleich mit frühabgesetzten Ferkeln zwischen Käfig- und Bodenhaltung. Züchtungskunde, 46: 458-467.

Te Brake, J.H.A., 19 . An assessment of the most profitable length of lactation for producing piglets of 20 kg body weight. Livest. Prod. Sci., 5: 81-94.

Troxler, J., 1979. Eine Alternative zur Flatdeckhaltung von Ferkeln. Vortrag gehalten an der Informationstagung "Tiergerechte Nutztierhaltung" für die Internationale Gesellschaft für Nutztierhaltung (IGN), 31 October, Basle, Switzerland.

DISCUSSION

D.L. Ingram *(UK)*

Were the younger animals, the litter as well as the sow, also inactive for most of the time?

R. Mauget *(France)*

The litter exhibited the same activity patterns as the sow. There was litter activity around the sow's bedding site, but only to a small extent.

D. Smidt *(FRG)*

You mentioned the function of the social group with respect to synchronisation of reproduction. What is the variation in the time of reproduction within these social groups?

R. Mauget

There is a close synchronisation of one week to ten days for reproduction within the social group.

D. Smidt

This means that they must have been mated within one week, more or less. What do you think are the physiological roots for this synchronisation - is it physical stimulation, or is it the boar? This will be significant when we come to discuss domestic pigs later on

R. Mauget

I do not know.

I. Ekesbo *(Sweden)*

Your Figure 2 shows bedding and feeding areas of radio-tracked animals. In Figure 5 you show the time budget. Have you made any observations as to whether the animals have special areas for urinating and eating? If so, have you taken into

G. van Putten

I do not know of any publication dealing with genetic differences. I think there are genetic differences but I am not very optimistic about the possibility of selection for a less susceptible piglet.

There are, of course, alternatives to straw. The best alternative that I know is woodshavings, but one must be careful since although some woodshavings make excellent bedding, one must not use tropical wood. The big disadvantage is that the shavings often contain a lot of sawdust which makes the pigs cough. There are equivalents to straw but I do not know if it is worth looking for these equivalents since there are huge amounts of straw available.

A.T. Smith *(UK)*

In your alternatives you did not mention the use of kennels of any kind - with straw, with insulated floors, and with slatted dunging areas outside. These would seem to be perfectly practical alternatives to flat decks, certainly in terms of piglet performance. Are there any reasons for not taking these into account?

G. van Putten

I am not against kennels at all. I mentioned this open front housing system which is doing very well with piglets. I only mentioned three alternatives, one of which was the pen with straw, the second was the open front - deep litter system. If straw or an equivalent material is available then I am sure that kennels will do as well as any other system.

R. Moss *(UK)*

Then are you rejecting a kennel used in conjunction with a partially slatted floor outside the actual lying area itself?

<u>G. van Putten</u>

The slatted floor is always very difficult, especially when the animals are growing as fast as piglets do; their claws grow very rapidly in this period. If the slats are very narrow then the piglets become very dirty, and if they are broader then, to begin with, their feet slip between the slats which causes wounds on the upper sides of their claws and on their heels which could cause infection of the legs. Slatted floors must supply two contradictory needs: they should give a large amount of safe room to stand on and also provide as much open space as possible to let through the dung. So, a compromise is necessary and in many cases the compromise is not working very well.

ANIMAL WELL-BEING AND HOUSING SYSTEMS FOR PIGLETS

R.G. Buré
Institute of Agricultural Engineering (IMAG),
Wageningen, The Netherlands.

ABSTRACT

Experiments are described in which different housing systems have been compared with respect to aspects of well-being and economics. There are risks connected with increasing the intensiveness of our animal housing systems. The animal has problems in adapting. Behavioural research makes it possible to recognise the symptoms which are the consequences of these problems.

The behaviour of piglets has been studied in a straw barn compared with a battery cage. The technical comparison showed hardly any difference. In the battery cage changes in activity and abnormal behaviour - the consequence of an inadequate environment - appeared to coincide and occurred more often.

The conclusion seems to be warranted that the straw barn for weaned piglets can be considered an alternative to present housing systems.

This is true in respect of the technical data as well as behavioural criteria for animal well-being.

1. INTRODUCTION

Years ago it had already been recognised that risks were connected with the increasing intensiveness of our animal housing systems; Riemann and Peters (1974) said: "Continuously the animal is made more and more subordinate to technical development. It becomes a product that passes through certain phases of production. Enormous technical possibilities are thus created, but at the same time great danger for the animal arises". Today we are indeed being confronted with this danger. Despite the great advantages of intensive animal production there are disadvantages as well. New housing systems are designed and the animal has to adapt. Although our domestic animals are flexible, the adaptibility of an animal is limited, both physiologically and ethologically. In present intensive animal husbandry limits are exceeded in a number of cases. This results in greater stress, blocking of normal behaviour, signs of fear or pain and an increasing morbidity and mortality: a bad effect on animal welfare.

2. REDUCTION OF WELL-BEING

Lorz (1973) describes animal well-being as a situation in which the animal lives in harmony with the environment and itself, both physically and psychologically. This means that pressure on adaptability is not too heavy. When the well-being of the animal is affected pain and defence reactions can result. When these symptoms occur, animal welfare has definitely been affected and the limits of adaptibility have been exceeded. Externally perceptible patterns of behaviour are closely connected with internal processes. That is why the study of behaviour is a real help in these situations. Problems of adaptability can be recognised at an early stage by measuring changes in the frequency of behaviour patterns. In view of the function of behaviour (maintenance of the internal homeostasis, protection and maintenance of the species) it can serve as an indicator for the function of an animal in relation to the environment. Abnormal behaviour often indicates an inadequate environment.

According to Sambraus (1978) one can talk of a challenge to well-being in a housing system when many or frequent behavioural patterns cease or are greatly altered. In research on animal behaviour, behavioural patterns can be measured in a precise way and - when comparing systems - quantitative differences can give information about the acceptability of a housing system.

In Figure 1, the connection between behaviour and well-being is shown schematically. In this model environmental stimuli are received by the animal and analysed. After that those stimuli are compared with a certain norm. This norm can have a genetic origin or be the result of learning. When environment and expectation differ, behaviour is performed to alter the situation. If the gap between reality and expectation is too large a situation of conflict or frustration occurs and changes in behaviour, or abnormal behaviour, appear.

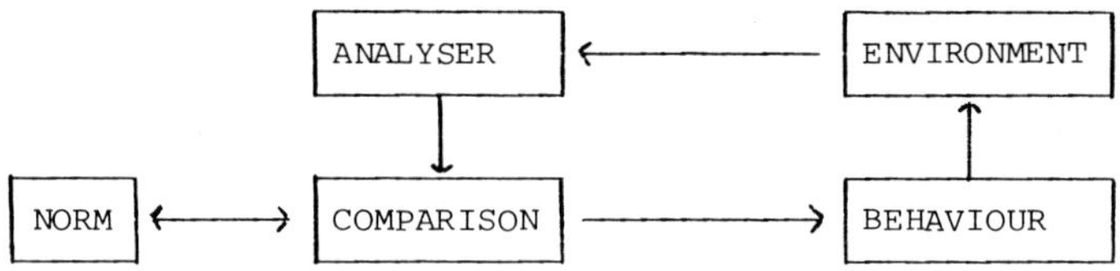

Fig. 1. Behaviour and well-being (Adapted from Wiepkema, 1980).

3. CRITERIA FOR ANIMAL WELL-BEING

When choosing criteria for animal well-being one must select from every behaviour an animal can perform.

Wennrich (1978) reports that an environment in which the animal cannot supply its wants can cause a situation of conflict, frustration or deprivation. In that situation the animal cannot bridge the gap between expectation and reality (Wiepkema, 1980).

According to Brantas (1978) in such situations the following symptoms can occur:

- changes in activity;
- use of substitutes;
- conflict behaviour;
- compromise behaviour;
- vacuum activities;
- stereotyped behaviour.

In present housing systems the above-mentioned behavioural patterns can be observed.

4. EXPERIMENTS

4.1. Material and methods

Alternatives to present housing systems in intensive animal husbandry have been designed and tested for some time. At the same time modifications are being made in existing housing systems to improve animal well-being.

In this experiment the open barn with straw for weaned piglets was compared with a battery cage system. The first system might prove to be a real alternative. After research on production (daily gain and feed efficiency), behavioural observations were carried out. The piglets were held in groups of 10 animals. The area/animal in the battery cages was 0.2 m^2. In the straw barns the animals were housed on an area of 0.5 m^2/animal. The air in the battery cages was heated and ventilated, while the air in the straw barn was not.

The pigsties in the straw barn were partly covered with an isolated cover. The piglets used the room underneath as lying area. If necessary the open side could be covered during the first days after weaning.

4.2. Results and discussion

Technical comparison of production data for both housing systems shows hardly any differences (Buré and Koomans, 1980), which justifies research into the straw barn (Table 1).

TABLE 1

AVERAGE FEED CONVERSION AND DAILY GAIN OF WEANED PIGLETS IN AN OPEN BARN WITH STRAW AND IN A BATTERY CAGE

	Open barn with straw	Battery cage
Number of animals	1775	1766
Average feed efficiency	1.79	1.75
Average daily gain in g	488	495
Animals culled	7	15

As has already been mentioned, behavioural research is directed towards measuring changes in behaviour and abnormal behaviour. If an animal has no opportunity to perform a specific behaviour or the expected result is not obtained, behaviour will be performed in an altered way, or another, often abnormal, behaviour will appear. With the aid of time-lapse video, the activity and resting behaviour of the animals was observed. Daily rhythms, measured in activity, can give quick information. Changes in activity can point at a challenge to well-being (Brantas, 1978). Difficulties with adaptation will occur sooner with behaviour that is genetically described. Because rhythm is genetic in nature, and genetics play an essential role in the adaptation of an animal to the environment (Baggerman, 1978), physiological and ethological disturbances can be an important indication of a challenge to welfare.

Figure 2 shows resting behaviour. An important difference between the two systems appears. In the straw barn there is a rhythm with two clear peaks of activity, while the battery

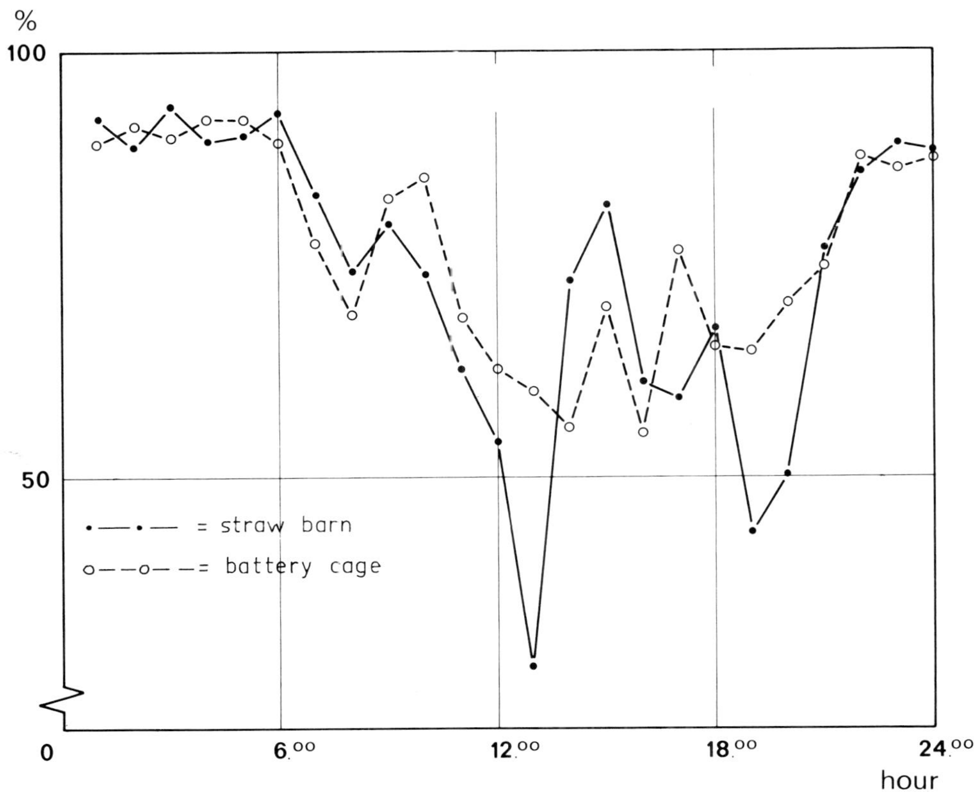

Fig. 2. Resting behaviour (amount of time spent resting).

cage shows an indistinct picture, especially in the afternoon and in the evening.

At the same time, abnormal behaviour was measured (conflict behaviour, redirected behaviour, vacuum activities). These behavioural patterns, mostly acts performed on other pigs, demonstrate that the environment is less suitable for the animal (van Putten and Dammers, 1976). Those patterns are given in Table 2.

TABLE 2

FREQUENCY OF ABNORMAL BEHAVIOURAL PATTERNS

Acts performed on other pigs	Open barn with straw	Battery cage	
	%	%	
Rooting	0.24	1.13	xxx
Nibbling	0.14	0.98	xxx
Massaging	0.56	1.09	xxx
Suckling	0.04	0.04	ns
Tail biting	0.01	0.14	xxx

ns not significant

xxx $P < 0.005$

Compared with the straw barn, both changes in activity (rhythm) and more abnormal behaviour can be observed in the battery cage. The more normal picture of the rhythm, more social influence and less abnormal behaviour indicate that the straw barn is for different reasons a more suitable housing system for weaned piglets.

Figure 3 gives the graph of abnormal behaviour, in an attempt to find a connection between that and resting behaviour.

There is a connection between resting behaviour and acts performed on other pigs (the correlation in the straw barn is -0.67 and in the battery cage it is -0.69) and the great

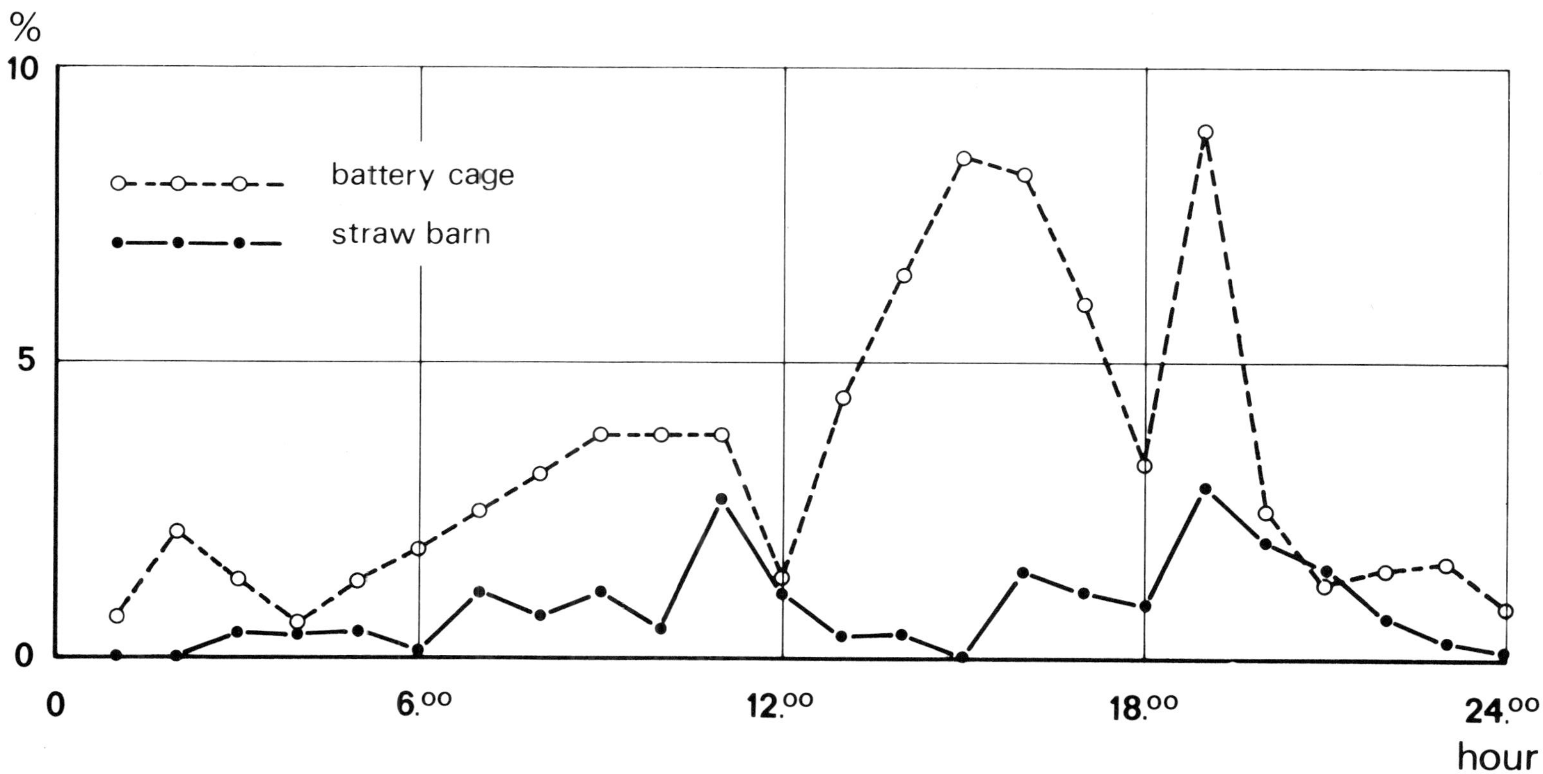

Fig. 3. Abnormal behavioural patterns.

difference in resting behaviour between the two systems coincides with the more abnormal behaviour in the battery cage. The connection between the rhythm of resting behaviour and abnormal behaviour underlines the possibility of using rhythm as a criterion for animal well-being.

5. CONCLUSION

Measuring well-being through behavioural observation is one of the methods of research into animal well-being which may lead to the enacting of laws on animal welfare. Only a combination of physiological, ethological and veterinary results can give a good, complete picture. In defining the acceptability of a housing system technical results, labour needs and building costs must be taken into account. This can lead to a form of animal husbandry that is fully justified.

REFERENCES

Baggerman, B., 1978. Ritmiek. In: Dijkgraaf en Zandee, Vergelijkende dierfysiologie. Bohn, Scheltema en Holkema.

Brantas, G.C., 1978. In: Sambraus, H.H., Nutztierethologie. Das Verhalten landwirtschaftlicher Nutztiere - Eine angewandte Verhaltenskunde für die Praxis. Verlag Paul Parey, Berlin-Hamburg.

Buré, R.G. and Koomans, P., 1980. Welzijn en huisvesting van varkens. De Boerderij 64 5 maart.

Deckert, R., 1968. Untersuchungen über die Leistungen und das Verhalten saugender Sauen bei Anbindehaltung. Diss. Deutsche Akad. der Landwirtschaftswiss. Berlin.

Lorz, A., 1973. Tierschutzgesetz. Beck'sche Verlagbuchhandlung, Munich.

van Putten, G. and Dammers, J., 1976. A comparative study of the wellbeing of piglets reared conventionally and in cages. Appl. Anim. Ethology, 2.

Riemann, U. and Peters, U., 1974. Entwicklung der Organisation von Tierbeständen. In: Comberg and Hinrichsen, Tierhaltungslehre. Verlag E. Ulmer, Stuttgart.

Sambraus, H.H., 1978. Nutztierethologie. Das Verhalten landwirtschaftlicher Nutztiere - Eine angewandte Verhaltenskunde für die Praxis. Verlag Paul Parey, Berlin-Hamburg.

Wiepkema, P.R., 1980. Gedrag en welzijn. Inaugurele rede Landbouwhogeschool Wageningen.

DISCUSSION

S. Baxter *(UK)*

Can you give us some information about the temperature regimes in the two systems - in the straw barn and in the flat deck cages?

R.G. Buré *(Netherlands)*

The temperature in the flat deck cages is automatically controlled and therefore constant. In the straw barn there are great differences especially when summer and winter temperatures are compared, but this does not worry the animals since there are individual covers. It is also possible to make a cover at the front so that the lying area in the straw is like a small house for the animals. They can create a sort of micro-climate in this covered part of the straw barn, so although in winter it might be freezing outside, it will be above zero within the covered part of the straw barn. In winter most animals lie under these covers and in the summer they lie in the open.

D. Smidt *(FRG)*

Varying influences within the systems may be the reasons for the differences found. The amount of space for the animals varied between the two systems. Do you think that some of the observations made could be on account of this difference?

R.G. Buré

First, I must say that the results from the observations are not, in my opinion, only due to the presence or absence of straw. The whole system has been studied. The lower building costs in the straw barn allowed more space per animal than in the flat deck cages. The main aim of our institute is to make observations on systems that are practical; we are comparing here two practical systems at the same price to see how big the differences are. We are not making fundamental research into the use of straw or the amount of space per animal.

J.P. Signoret *(France)*

I don't altogether agree with your approach. For example, surface per animal plays a critical role in the frequency of behavioural interaction. If it is possible to have a well insulated floor without straw, it will change the price in relation to the cost of the straw, the possibility for mechanisation of cleaning, etc. So, I think it would be useful to undertake some kind of experiment dealing with the influence of surface per animal.

R.G. Buré

I do not make investigations into the influence of one specific factor on the whole animal behaviour because this is not the first priority of my institute. I would like to but it is impossible at the moment. At present we are simply comparing two systems, the battery flat deck cage and the straw system. Next February, we will make the next comparison - the open barn with straw with the partially slatted floor. Finally we will be able to put the results together and show which is the most important factor: use of straw, type of floor, climate, food, etc. Observations from many different systems may provide the answer on a particular question regarding a certain factor effect.

SHORT COMMUNICATION

R. Moss *(UK)*

I believe that Professor Adler, before he gives his own paper, has something to add to the common pool of knowledge on this particular matter of slats and straw.

H.C. Adler *(Denmark)*

I want to give some information from Danish observations. In Denmark, it is the Ministry of Justice which deals with animal welfare matters. Last year this Ministry published a report on intensive production systems and animal protection in practical agriculture in Denmark based on literature and visited production systems. I would like to give you some information from the report of the swine section in relation to the question of slatted or solid floors, and I have chosen the tail-biting problem to illustrate what can be found in this report.

Altogether 4 500 fattening pigs were observed. They were on solid floors with straw and slatted floors, respectively. It was found that the risk of tail-biting was 4.5 times as high on slatted floors as compared to solid floors with straw. It was also found that tail-biting was a much earlier developed phenomenon in the life of the pig on slatted floors as compared to solid floors. If two-thirds of the slatted floor was covered by some solid material you could reduce the frequency of tail-biting significantly. Furthermore, it has been calculated that tail-biting is responsible for 14% of the total loss from condemnation in Danish slaughterhouses. So, here we have an example of a very solid economic reason for considering various raising systems.

DANISH STUDIES ON BEHAVIOUR OF EARLY WEANED PIGLETS: PRELIMINARY RESULTS

Mette Schmidt and H.C. Adler
Royal Veterinary and Agricultural University,
Copenhagen, Denmark

1. INTRODUCTION

The aim of the studies was to evaluate the effect of early weaning and housing in cages on the behaviour of piglets and to compare 2 stocking densities.

2. MATERIALS AND METHODS

Three experiments were carried out. For each experiment 3 littermate gilts were served by the same boar. Within 24 h after farrowing, the piglets were given numbers which were painte on their backs and tattooed in one of their ears. Two litters were weaned at 3 weeks of age and placed in cages where the floor area was 0.20 m^2 and 0.15 m^2/piglet, respectively. One litter served as control in which the piglets stayed with the sow in a pen with straw until 8 weeks of age.

At 8 weeks of age all 3 litters were moved to pens with straw.

The experiment was repeated 3 times but because of too few piglets in one litter, the 0.20 m^2 litter was excluded in one experiment. Thus the total number of litters observed was 8, containing a total of 62 piglets.

Each litter was observed 4 x 1 h/week during the third, fourth, sixth, eighth and ninth week of age.

For each successive minute during the hour of observation the behavioural activities of each piglet were recorded. The number of minutes in which a piglet showed a given activity was defined as pig-minutes. To compare the behaviour between litters of different size and in which different numbers of piglets were asleep, the number of pig-minutes for an activity was expressed as a percentage of the number of piglets awake during the hour of observation.

The following behavioural elements were recorded:

1) Playing: which consisted of 2 activities:

   a) Playful fighting involving 2 or more piglets with no aggressive elements.

   b) Running around, finishing the quick run by throwing themselves on the floor or against another piglet. This activity sometimes induced playful fighting.

2) Aggression: which involved only 2 piglets and which consisted of real fighting, bites and pushes.

3) Exploration: which consisted of rooting with the snout or pawing with the front hoof over the floor and nibbling on parts of the cage with the mouth.

4) Sucking: which was a chewing, nibbling and sucking upon ears, tail and other parts of the body of another piglet.

5) Belly-nosing: which was an up and down massage with the snout and back of the nose on the belly and sides of other piglets.

2.1 Statistical analysis

The effect of age on behaviour was analysed by χ^2-tests, pairwise comparing the sum of pig-minutes each week of observation with those of the following week.

The effect of the housing system on behaviour was analysed by student's t-test for pairwise comparisons.

3. RESULTS

The number of pig-minutes as a percentage of piglets awake is given for each of the activities in Figures 1 - 5.

From these figures the changes of the behaviour with age, and the influence of the housing system can be observed.

3.1 Age

An outline of the changes with age is given below:

TABLE 1

		In the 4th-8th week of age	9th week of age
Playing	cages:	Decrease after weaning	Increase
(Fig. 1)	control:	Increase with age	Decrease
Aggression	cages:	Increase after weaning	Decrease
(Fig. 2)	control:	Slight increase with age	Increase
Exploration	cages:	Increase after weaning	Decrease
(Fig. 3)	control:	Slight increase with age	Increase
Sucking	cages:	Increase with age	Decrease
(Fig. 4)	control:	No change with age	Increase
Belly-nosing	cages:	Increase with age	Decrease
(Fig. 5)	control:	No change with age	Increase

The early weaned piglets showed after weaning at 3 weeks of age a high level of aggression, exploration, sucking and belly-nosing. Furthermore, the sucking and belly-nosing increased with age. The playing was the only one of the observed activities which showed a decrease after weaning and a low level in the cages. When the piglets at the age of 8 weeks were removed from the cages the reversed picture was seen - the aggression,

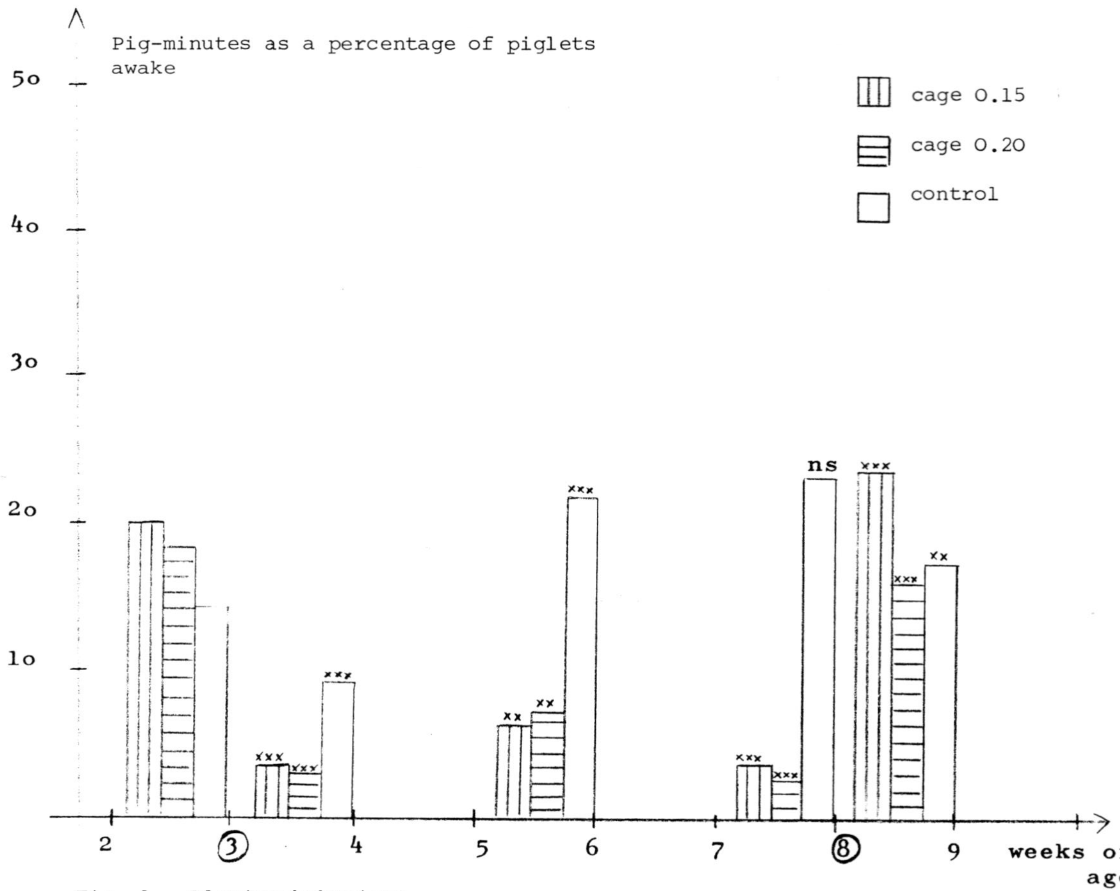

Fig. 1. Playing behaviour.
P < 0.05 : x; P < 0.01 : xx; P < 0.001 : xxx; for significant differences between results from one week and the preceding week.

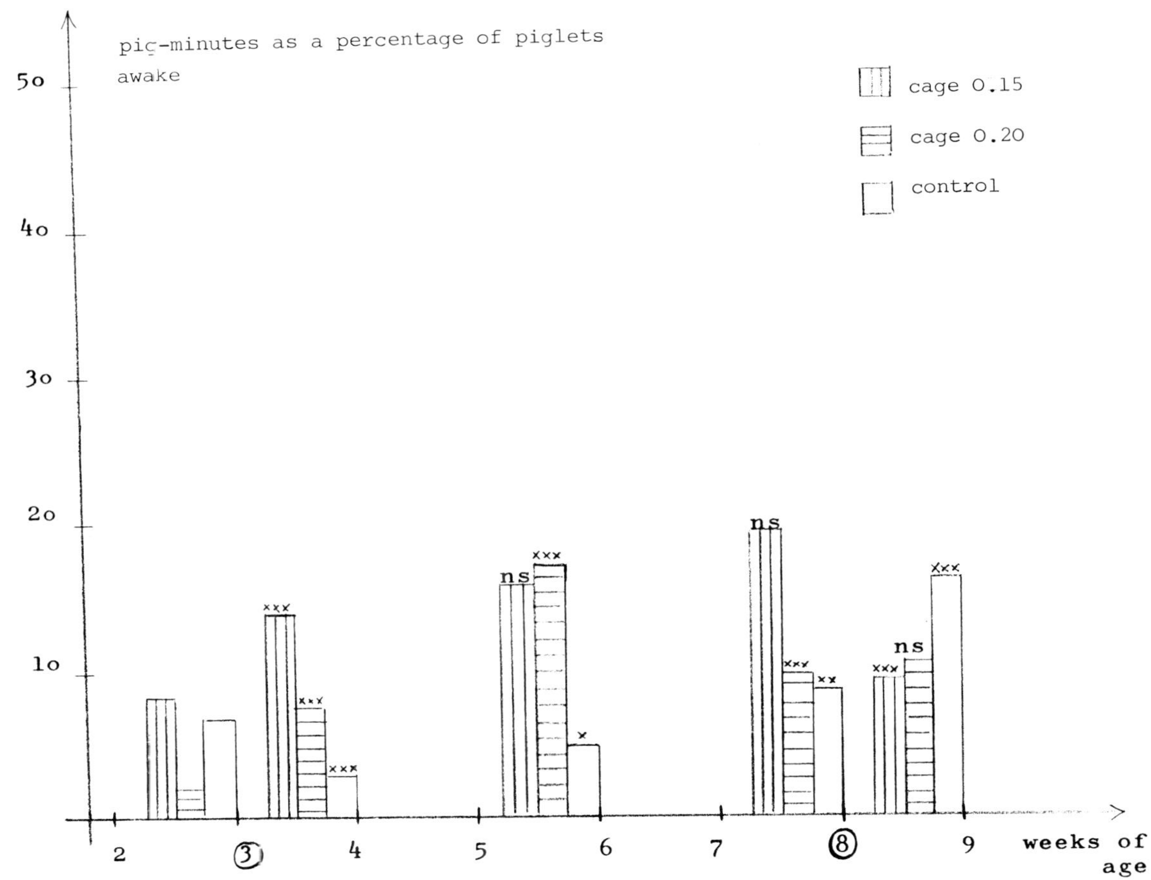

Fig. 2. Aggression
P < 0.05 : x; P < 0.01 : xx; P < 0.001 : xxx; for significant differences between results from one week and the preceding week.

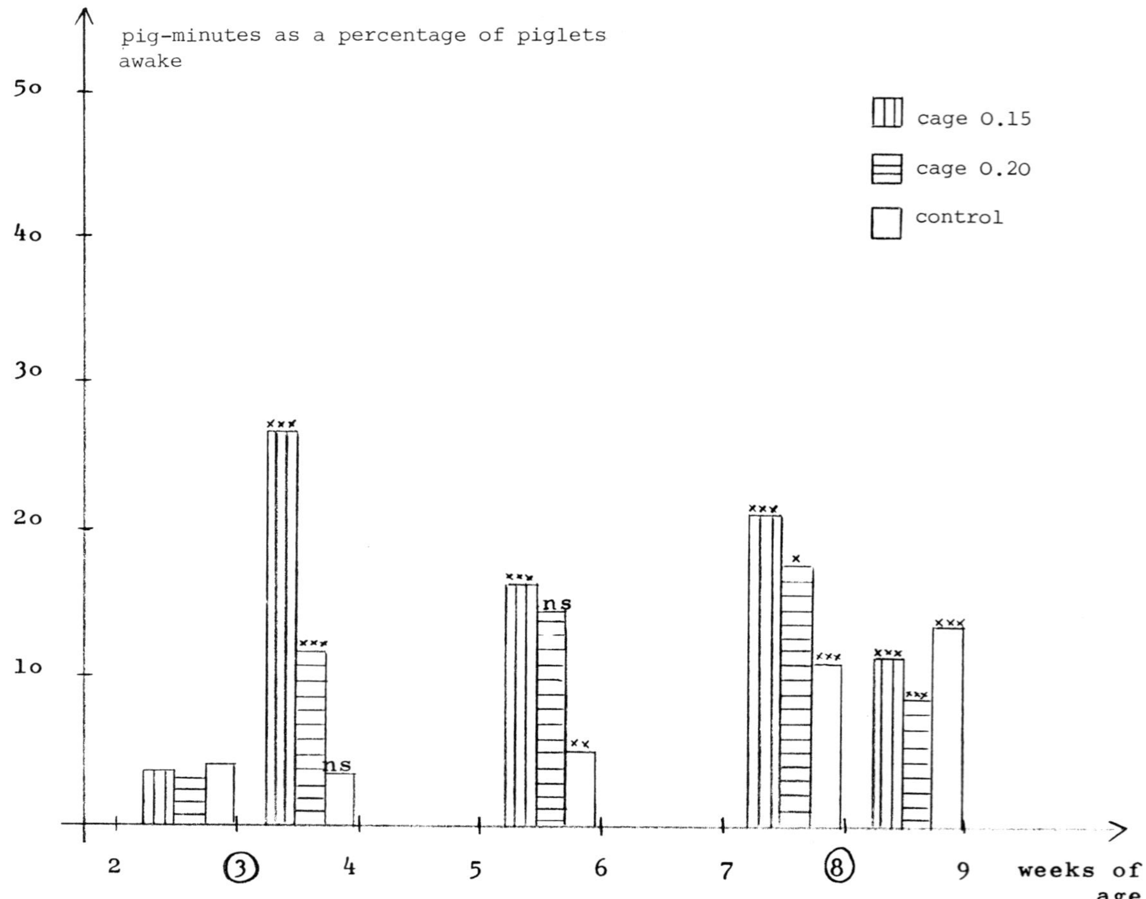

Fig. 3. Exploration
P < 0.05 : x; P < 0.01 : xx; P < 0.001 : xxx; for significant differences between results from one week and the preceding week.

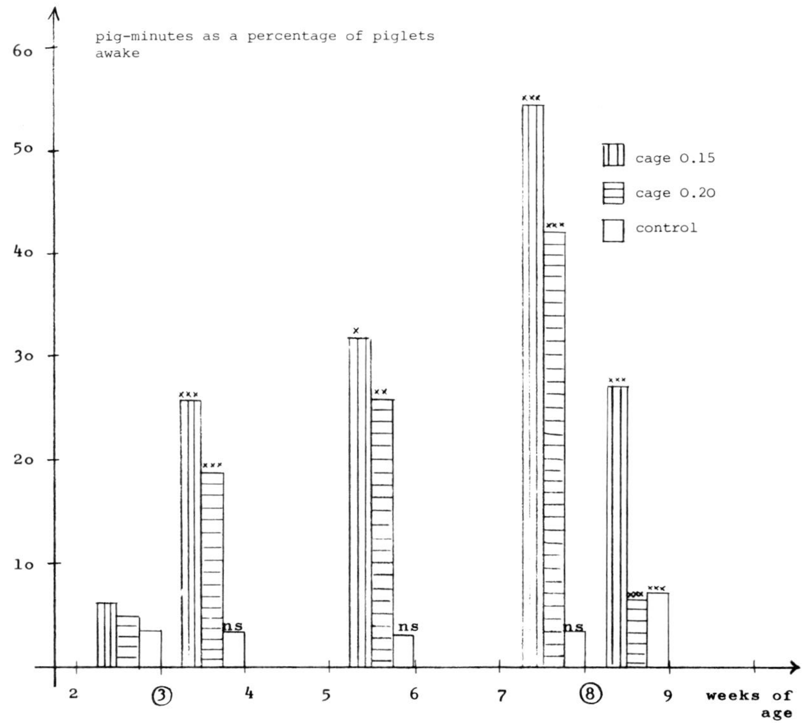

Fig. 4. Sucking
P < 0.05 : x; P < 0.01 : xx; P < 0.001 : xxx; for significant differences between results from one week and the preceding week.

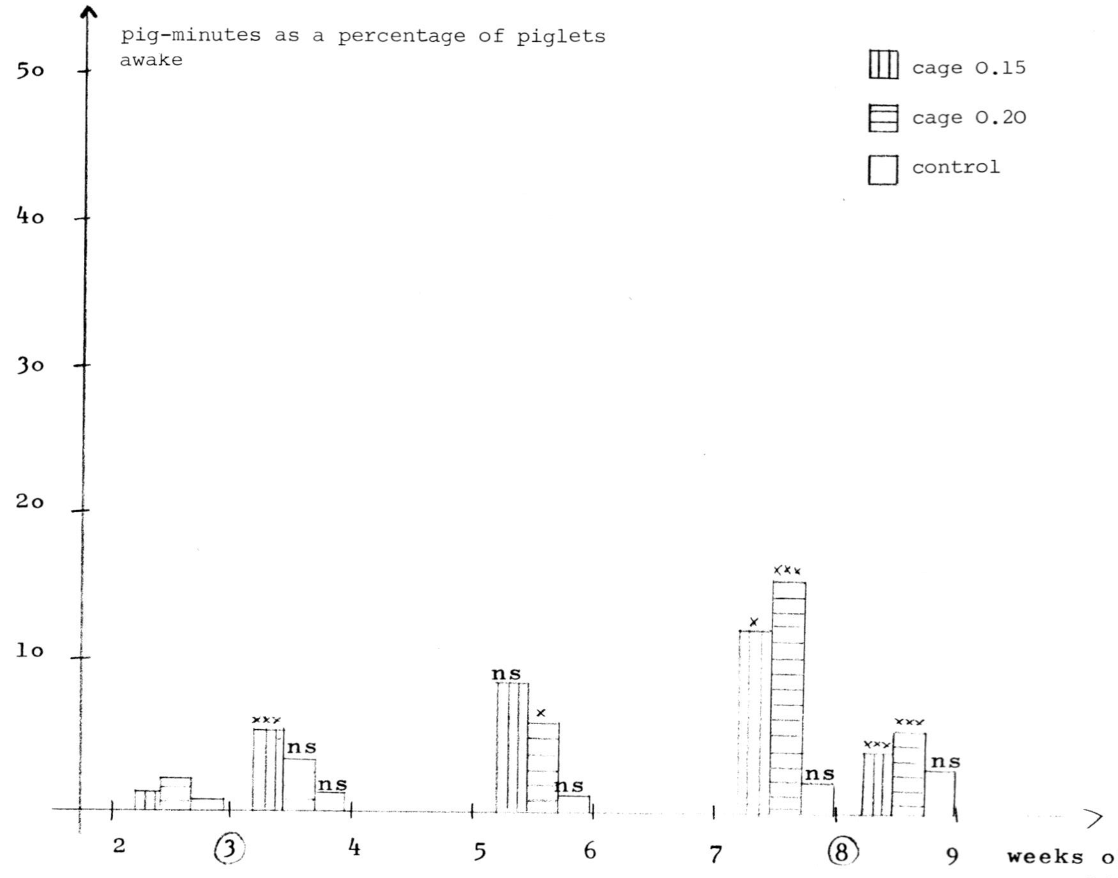

Fig. 5. Belly-nosing
$P < 0.05$: x; $P < 0.01$: xx; $P < 0.001$: xxx; for significant differences between results from one week and the preceding week.

TABLE 2

BEHAVIOURAL DIFFERENCES IN RELATION TO HOUSING DIFFERENCES

		0.15 m^2/pig	0.20 m^2/pig	control
Playing (Fig. 1)	4th-8th week	(less than →	→	xxx
		(	less than →	xxx
	9th week	more than →	x	
Aggression (Fig. 2)	4th-8th week	(more than →	→	xxx
		(	more than →	xxx
	9th week	less than →	→	x
Exploration (Fig. 3)	4th-8th week	(more than →	→	xxx
		(	more than →	xxx
	9th week			
Sucking (Fig. 4)	4th-8th week	(more than →	→	xxx
		(	more than →	xxx
	9th week	(more than →	→	x
		(more than →	x	
Belly-nosing (Fig. 5)	4th-8th week	(more than →	→	xxx
		(	more than →	xx
	9th week			

$P < 0.05$: x; $P < 0.01$: xx; $P < 0.001$: xxx

the exploration, the sucking and the belly-nosing decreased and the playing increased.

The control piglets, while staying at the sow, showed a slight increase with age of the play, aggression and exploration, while sucking and belly-nosing did not change with age. After weaning at 8 weeks of age there was a decrease in playing while other behavioural activities increased.

3.2 Housing system

An outline of the behavioural differences is given in Table 2.

During the fourth to eighth week of age, both kinds of cage piglets played less than the control piglets and during the ninth week of age, the 0.15 cage piglets played more than the 0.20 cage piglets.

During the fourth to eighth week of age, both kinds of cage piglets showed more aggression, exploration, sucking and belly-nosing than the control piglets. During the ninth week of age, the 0.15 cage piglets showed less aggression and more sucking than the control piglets. Of the 2 kinds of cage piglets, the 0.15 cage piglets showed most sucking.

4. DISCUSSION

The results of this behavioural study were, for the most part, in agreement with results obtained by other investigators of early weaned pig behaviour (van Putten and Dammers, 1976; Marx, 1977; Schmidt, 1977; Fraser, 1978). An exception was playing behaviour and the main reason for this is the definition of playing behaviour. In the present study, early weaned piglets in cages played significantly less than the control piglets. Van Putten and Dammers (1976) found no differences between the same kinds of groups. The variation between these two results can be explained by the fact that van Putten and Dammers, in their definition of play, included aggressive fighting which increases with early weaning and thus compensates for the decrease of play. Marx (1977) found even a higher frequency of play in the cages than in control piglets. An explanation of this result is that in his definition of play he included biting in and sucking on the cage and other lifeless objects. These activities in fact increase with early weaning (van Putten, 1978; Schmidt, 1977) but in our investigation they are defined as part of exploratory behaviour and by van Putten (1978) as conflict behaviour.

Play could be taken as a measure of the well-being of the piglets. After weaning which is considered to be a stressor this behaviour decreased and when the cage piglets left the cages, which were small and without adequate stimuli, the play increased.

The increase in aggressive and exploratory behaviour with early weaning is in agreement with the results of Fraser (1978). It could be a result of discomfort or of gastrointestinal malfunction and anxiety (Stanton, 1976). The high stocking rate in the cages could also give rise to many aggression interactions. The lack of straw or other suitable materials could be the reason for the high level of exploratory behaviour in the cages.

The high level of aggression which is seen in the cages as well as in the control litters after weaning should be taken as an indicator of a low level of well-being. This assumption is confirmed by the fact that the aggression decreased when the piglets left the cages.

The high level of sucking and belly-nosing in the cages was also found by van Putten and Dammers (1976), Schmidt (1977) and Fraser (1978). Since these abnormal oral behavioural activities were repeated regularly and frequently they should be considered as stereotypes which developed because of the lack of any possibility of suckling the sow at an age when the piglets were highly motivated towards this kind of behaviour. Due to its stereotypical nature, this abnormal oral behaviour did not disappear when the piglets left the cages, although it did decrease.

The theory about the behavioural alteration as indicators for a stressful situation is confirmed by Worsaae and Schmidt (1980) who found that the decrease in playing and the increase in aggression, sucking and belly-nosing in early weaned piglets were correlated with an increase in plasma cortisol.

5. CONCLUSION

The conclusion of this study is that the early weaning of piglets and then the rearing of them in cages are stressors which change normal behaviour and give rise to the development of abnormal behaviour.

REFERENCES

Fraser, D., 1978. Observations on the behavioural development of suckling and early weaned piglets during the first six weeks after birth. Anim. Behav., 26: 22-30.

Marx, D., Schrenk, H.-J. and Schmidtborn, C., 1977. Spiel- und eliminationsverhalten von Saugferkeln und frühabgesetzten Ferkeln in Käfiggruppenhaltung (flatdecks). Dtsch. Tierärztl. Wschr., 84: 125-164.

van Putten, G. and Dammers, J., 1976. A comparative study of the well-being of piglets reared conventionally and in cages. Appl. Anim. Ethol., 339-356.

van Putten, G., 1978. Das sogenante Spiel der Flatdeckferkeln. Dtsch. Tierärztl. Wschr., 85: 457-496.

Schmidt, M., 1977. Svins adfaerd ved tidlig fravaenning. Dissertation, Royal Veterinary and Agricultural University, Copenhagen, Denmark.

Stanton, H.C. and Mueller, R.L., 1976. Sympathoadrenal neurochemistry and early weaning of swine. Am. J. Vet. Res., 37: 779-783.

Worsaae, H. and Schmidt, M. Plasma cortisol and behaviour in early weaned piglets (in press).

DISCUSSION

<u>B.W. Moss</u> *(UK)*

In your last Table which gives aggression differences you show that there was less aggression where there was less space per pig. Is that because there was not the physical space to allow aggression to take place? Could this be a possible reason for those observations given that normally if you decrease the space per pig you expect more aggression?

<u>H.C. Adler</u> *(Denmark)*

This may have something to do with it.

<u>B.W. Moss</u>

Secondly, in relation to cortisone measurements, was there any difference between the two space allocations?

<u>H.C. Adler</u>

I cannot tell you at the moment, we are in the process of analysing the data.

<u>K. Vestergaard</u> *(Denmark)*

I have a comment about playing behaviour. Some people find more play and some people less play, in cages. We need some basic studies to establish exactly what is playing. If you include nibbling on objects etc. then you will find more playing in cages. So, if playing is taken as a measure of well-being, it is best in cages. If you use other criteria you find the best well-being in the pen with straw. This is a problem of interpretation of such comparison studies.

<u>B.W. Moss</u>

In relation to the last comment, it is important to interpret the results and consider whether the play activity may lead eventually to other activities such as aggression.

From this viewpoint, it is important to interpret the basic behaviour studies in relation to the well-being of the animal.

GENERAL DISCUSSION

R. Moss *(UK)*

Since yesterday afternoon we have been looking at the way in which we keep fattening pigs, piglets and sows, and discussing the reaction of those animals to the environments that we provide for them. The basic thing that we provide is space, not just the square metres on which the animal sits, but we provide space in all that space implies. We are looking at the quality of that space. We have looked at floor types, at fittings. With regard to floor types, we have only talked about slats but we have seen a number of slides dealing with other types of floor - punched metal floors etc. Are they any different to slats? Have they any influence on animal welfare?

Feeding systems: we discussed a feeding system yesterday where it is possible that the amount of trough space allowable had an influence on the behaviour of the pigs. I think there is a need to discuss in depth whether or not straw should be used.

We need also to agree on measurements. Health; diseases and injuries that we see. Production; maximum production does not necessarily mean good welfare. Behaviour; I would like to hear views about the stimulus that the space that we provide gives to the animal. How many stimuli are there? How many are necessary? What happens when those stimuli are too many, or too few? How do you measure that change in behaviour? Do you start off with the base line that the measurement is the departure from the range which is normally expected and observed? When we describe behaviour, I think that we need to be extremely careful, and I would like to have descriptions of behaviour. A moment or two ago, we had an interesting comment about play. It is easy to use the word 'play' because when one scientist uses the word to another, or even to a layman, there is an immediate impression of what is going on. It looks like play, but it may not be play within the animal's repertoire. There are other descriptions that I think sometimes we use very loosely and very dangerously. Does a departure from what is expected behaviour mean that the animal is suffering? How do we know? Are there, in fact, any essential behavioural needs that we need to provide within the space in which we confine the pig? If so, which are they?

I think that physical health goes, without saying, as the baseline on which we then consider all the other measurements. But how do we prove any of this? I would like to come back to the point I raised yesterday afternoon that we should begin to look from the inside out, rather than from the outside in, in an attempt to consider the consciousness of the pig and how it perceives the world and the space that we provide for it.

I ask for comment and questions. I know that Dr. Truyen has wished to put a question about another paper, and perhaps he might like to do so now.

W.T. Truyen *(Netherlands)*

I believe that weaning will occur between 4 and 5 weeks, and the piglets will go to fattening houses at about 10 to 11 weeks and not 8 to 9 weeks as Dr. van Putten stated. We must take this into account when using weaning houses, that we need them for 5 to 6 weeks and not 3 to 4 weeks.

If we use straw in the weaning houses, I believe that we will lose quite a few piglets due to crushing by the sow. Perhaps we ought to look for some better bedding.

G. Thines *(Belgium)*

I would like to make a comment about space. I think it is true that it is a very important topic. However, what space means to the animal cannot be given an absolute measurement. Space is a subjective definition and can only make sense if you know, first, what is the normal range of perception for a given animal, and secondly, how it is used, in actual behaviour. Therefore, one difficulty in assessing what space is needed is the lack of knowledge about the behavioural importance of the range of senses in domestic animals. This can give the basis of interpretation of what an animal needs.

Another important point emerges when you work with groups. I think that it is quite senseless to think that if you know the mean range for a given individual of a species that you may consider that the number of animals will give you a linear function of the needed space. It is certainly not in all cases a linear function. If one is dealing with 5, 10, 20 or 100 animals, according to the species, I do not think one can say that the space should be multiplied by 'x'. This should be qualified by knowledge of the specific range. This is an important aspect of applied ethological problems, in relation to the problems of housing, etc.

R. Moss

I think that you have made a very important point that the baseline depends upon what the animal is doing at the time. It may require very little, it may require a lot. When we say require, how do we know?

J.P. Signoret *(France)*

Turning to one of the problems that has been discussed this morning, that of play; I was very much impressed by the convergence of the results that showed that there was less play and more abnormal behaviour on slatted floors compared with piglets kept on straw with the sows. However, I am very surprised in comparing such results with what we have observed with lambs. With artificially reared lambs, kept only among young animals, either on slatted floors or straw, the importance of play is increased 5 times when compared to the play when the lambs are kept with their dams. It can hardly be concluded that the lambs are happier when separated at birth from their dams. To conclude, we need experiments under practical conditions, as we have to answer to farmers and lawyers.

Doris Buchenauer *(FRG)*

I would also like to make a comment about playing. In my opinion, playing is one of the best described behaviour patterns in the literature. I think that if the investigators carefully studied the literature before they start experiments, there would be much less misunderstanding than there is at the moment. Interpretations for intensive environments are also clearly described. It is a very intensive parameter, and I would like to ask Dr. Buré how playing and locomotory patterns appear in his experiments.

R.G. Buré *(Netherlands)*

The things that I have shown today are some of the behavioural patterns that I have observed. In my investigations playing behaviour as described in the literature occurs much more in straw barns than in battery cages. I think that especially with young piglets, it is very difficult to make a division between playing and biting behaviour. That is why in my situation I have combined them. The combination of playing and fighting - fighting as pure aggression, not cannibalism - occurs more in the straw barn than in the battery.

R. Moss

We are using the word 'play' in the meaning that it is a happy thing. The function of play is a learning function, is it not, and is that necessarily a happy situation. Perhaps Dr. Dantzer would comment on this.

R. Dantzer *(France)*

I must admit that I have not considered this question so I cannot answer it in any very direct way. However, I would like to point out that we can deal with the problem of looking at different management systems in two ways. Firstly, there is the way which simply describes the differences that occur at the time. This is important but it is not enough. If you are expecting the system to have some influence on the animal, this must be seen not only as an immediate influence but also as a long term one. We have tried to see the long term effects when dealing with 7 week old weaned piglets, and see the way the animal reacts to over-experience. You would expect that if animals are suffering from their conditions they would display some deficit. However, we have done this sort of experiment but have been unable to find any deficit in pigs that have been submitted to different experiences during their early life. Does this mean that our tests are inadequate, or that pigs have good adaptive abilities?

Doris Buchenauer

I would like to make a further comment to Dr. Buré. I do not find it surprising that the animals fight more; it is probably what is described in the literature as fighting-playing, in which they do not hurt each other. I think that it is an important positive behavioural pattern because it is part of their social education.

In regard to Dr. Signoret's comments; I have also observed the playing behaviour of lambs. When lambs are with their mothers on the pasture, in summer, we observed fewer behavioural patterns, far fewer than at the same period in the stable with the artificially reared lambs. In the autumn we observed many more behavioural patterns on the pasture by the lambs with their mothers. This shows that playing behaviour depends to a large extent on environmental factors like climate - sunshine, heat, etc. I do not think it true to say that motherless lambs play more than those raised with the mother.

J.P. Signoret

I would just like to comment on the possible learning effects of playing with regard to lambs rather than pigs. We could hypothesise that the sexual patterns involved in the play of young lambs could have a learning effect on later sexual behaviour. In fact, in a recent experiment, we have found that there is absolutely no difference in the development of adult sexual behaviour, in regard to whether lambs have had this sex-play or not.

K. Vestergaard *(Denmark)*

I have a comment on the long term effects of treatment. There is at least one study which shows that if pigs are handled before weaning and stroked gently a few times each day, this increases aggression much later in life.

G. van Putten *(Netherlands)*

Mr. Chairman, you posed some questions concerning behaviour. I think that you are more convinced by the veterinarian data than you are by the behavioural data. I think that most of the audience here thinks likewise. If the animal is ill or injured then this is a deficiency of well-being that can be clearly observed. If we really tried to discuss the validity of the behavioural observations, I think that we would need at least one week. However, perhaps we could come to a conclusion based on the opinions of the specialists here instead of debating all sorts of definitions and validities. I think that some things have become clear from the papers already given. I think that most of us here agree that there has to be some furnishing, which could be straw or an equivalent of straw which would give the animals the opportunity to do something. They should not be idle all day except for the short time they need for feeding.

B.W. Moss *(UK)*

Mr. Chairman, in the comment that you made earlier, you said that it was important to discuss how the pigs perceive their environment and we had another comment in this direction in the interpretation. Yesterday, Dr. Ingram commented that from the welfare point of view maybe the best approach is to end up with conditions that please the animal welfarists. I wonder if these two, considering the perception of the pigs and keeping the animal welfarists happy, are maybe on similar lines. In our interpretation of both of these we are attributing human feeling to animals. Perhaps this could be illustrated in relation to the space per pig, as this seems to me to be one factor in intensive situations that the general public would pick on. Should we create an environment where we can see that there is ample space per pig - not the actual space per pig, but the space occupied by the pig? As Dr. Baxter mentioned earlier on, this could be changed by temperature and other environmental conditions. I suggest we put the idea of perception and the idea of keeping the human attributes of animal welfare together, and produce a situation which is pleasing to both, and would give a general impression of well-being.

D.L. Ingram *(UK)*

I am rather surprised that no research seems to have been done on the shape of the area in which an animal is kept. I do not mean circular or square, but the possibility of providing some central barrier so that instead of the animal being able to see his whole environment he has to explore it by going round something. If this is done to mice then you can change their sort of activity. I wonder whether animals could be given the illusion of having more space simply by changing the shape of the enclosure.

J. Daelemans *(Belgium)*

Mr. Chairman, you posed the question whether or not to use straw. If we look further into the disadvantages and advantages of straw, there are many disadvantages of straw to be found. Firstly, there is a more complicated situation for the stockman. I made some calculations for the cereal importing countries such as Belgium and the Netherlands. If there is a move to use more and more straw it will be getting very expensive.

W.T. Truyen *(Netherlands)*

I would like to make one point as a veterinarian. Up till now we have been discussing systems for keeping piglets, with and without sows. I am afraid that we are only talking about healthy piglets. I think that we should also consider the situation in practice where we are often working with unhealthy situations. What is the influence of using straw or not from this point of view? For example, when confronted with diarrhoea caused by say Salmonellosis, dysentry etc., the building up of the infection in the straw bedding house will be faster than in the slatted floor or other meshed floor houses. I think that this is an important point to consider.

I. Ekesbo *(Sweden)*

I can mention, concerning salmonella, that we have made studies in Sweden on this very point. It is quite clear from all the investigations made that there is a significantly higher salmonella incidence in herds with no straw and liquid manure handling, than with straw and solid manure handling. Three different investigations have taken place in the last 10 years and all have shown the same results.

We have also had practical experience of the difficulty of disinfection in herds where we have liquid manure handling. In fact, we do not know how to deal with this problem, especially in respect to salmonella.

W.T. Truyen

It is not my aim to state a preference, my only aim is to stress that we should consider these points in the discussion; also, for instance, the building up of immunity in pigs. It is known that this is influenced by environmental conditions.

R. Moss

I must now bring this discussion period to a close. I have noted 3 particular points: first, I believe there is still a continuing need for basic research in the area of perception consciousness. The second point was that of the Council of Europe, and may I remind everyone here that the Convention for Keeping Farm Animals says that they shall be kept in accordance with their physiological and behavioural needs, in accordance with current practice and the latest scientific knowledge. Finally, we must not forget that there are social and economic consequences of the way in which we decide that we are going to keep livestock.

SESSION III

COMPARATIVE DATA REGARDING WELFARE IN PRESENT HUSBANDRY SYSTEMS, I.

I. FATTENING PIGS

Chairman: R. Moss

EXOGENOUS STIMULI AND REPRODUCTION OF THE SOW

A. Hoogerbrugge
Institute of Zoötecnic, Veterinary Faculty,
State University, Yalelaan 17, Utrecht, The Netherlands.

1. INTRODUCTION

Statistical figures from many countries make it clear that anoestrus and delayed oestrus are very serious problems in pig reproduction. These problems are most common in first-litter sows (Corteel et al., 1964; Rasbeek, 1969; Tuinte, 1976; Gordon and King, 1978; King, 1978; Bisperink, 1979; Love, 1979; Stork, 1979; Hurtgen and Leman, 1980; Karlberg, 1980; Hemsworth et al., 1980).

Physiological and veterinary research have not solved the problem. The main thing that is done in practice is injecting the female with sexual hormones. This injection is more for the farmer's benefit than the sow's because it is therapy aimed at the symptoms and not at the cause. It seems that male and female pigs need special stimuli for normal reproduction. The environments of our modern labour-saving piggeries are probably so oversimplified that a lot of pigs lack the basic stimuli for normal reproduction.

2. MANIFESTATION OF OESTRUS

Since the introduction of artificial insemination it is of great importance that the person involved can recognise whether or not a female is in heat. Besides change in the vulva, the investigations of Signoret et al. (1975) have shown that the female pig needs different stimuli before she will show the standing reflex (Table 1).

These findings are very important for practical pig-husbandry. Furthermore, tether boars and the so-called 'boar parfum' are used for oestrus detection.

TABLE 1

EFFECTIVENESS OF VARIOUS STIMULI IN RELEASING THE POSTURAL STANDING REACTION OF THE OESTRUS SOW (SIGNORET, 1975)

Stimulation	Postural standing reaction(%)
Pressure on back only	48
Broadcasting boar's courting grounds	70
Odour of preputial secretion at 40°C	80
Smell and sound of boar	90
Smell, sound and sight of boar	97

3. STIMULATION OF OESTRUS

3.1. Gilts

The gilts of modern breeds reach puberty at the age of about six months. Apart from genetic differences it is obvious that there are environmental factors which influence the age of first oestrus.

3.1.1. Transport effect

It was found by du Mesnil et al. (1962) that transportation of gilts from one pig farm to another stimulated the onset of puberty in these young females.

Besides this 'transport effect' different investigators have found a 'male effect', i.e. the presence of a boar can stimulate the onset of puberty. The most important stimulus of these two seems to be the boar effect. This was found by Bourn et al. (1975) and is indicated by the results shown in Table 2.

More complete information would be available if a control group could have been used to investigate the effect of transport on its own.

After these publications most research concentrated on the 'male effect'.

TABLE 2

INFLUENCE OF AGE, TRANSPORT AND BOAR EXPOSURE ON PUBERTY ATTAINMENT IN GILTS (BOURN ET AL., 1975)

Treatment	Age at puberty (days)
Gilts 135 days transport (2.8 km)	175.5
Gilts 165 days transport (2.8 km)	193.4**
Gilts 165 days transport + boar exposure	178.0
Gilts transport at 135 days and boar exposure at 165 days	173.7

** P < 0.005

3.1.2. Age of the gilts

Contact with the boar acts as a stimulus only when the gilts are not too young. This was found by Brooks and Cole (1969) and the results are shown in Table 3.

TABLE 3

EFFECT OF CONTACT WITH BOARS DURING THE REARING PERIOD ON PUBERTAL AGE AND WEIGHT OF GILTS (BROOKS AND COLE, 1969)

	A Isolated from boar in pre-experimental period	B Contact with boars from 122 days of age
Age at start of experiment (days)	171.1	187.7
Weight at start of experiment (kg)	83.1	84.4
Age at puberty (days)	192.1	234.1
Weight at puberty (kg)	90.0	99.7

These results would seem to warrant the conclusion that when gilts grow accustomed to the boar at too early an age there is no later boar effect.

3.1.3. Duration of exposure

Kirkwood and Hughes (1980) investigated the duration of exposure to the boar (Table 4).

TABLE 4

THE EFFECT OF BOAR EXPOSURE ON PUBERTY ATTAINMENT AND OVULATION RATE IN THE GILT (KIRKWOOD AND HUGHES, 1980)

	n	Age at puberty (days)	Time to puberty (days)	Ovulation rate
30 min. boar exposure	10	198.7	33.7	15.2
continuous boar exposure	10	194.6	29.5	15.0
control	10	232.1**	67.1	15.3

** $p < 0.01$

The overall conclusion is that 30 min boar exposure/day gives the same stimulus as a continuous boar exposure in inducing heat.

The possible components of stimuli given by boars may be visual, tactile, auditory and olfactory. At present it seems that visual and auditory cues and olfactory stimuli of urinary origin do not play a significant role as stimuli (Zimmerman et al., 1976; Kirkwood and Hughes, 1980). Recent publications by Hughes and Kirkwood (1980) emphasise this. They investigated the role of the odour of the boar (Table 5).

The conclusion to be drawn from this investigation is that the major stimulus in the 'boar effect' appears to be olfactory in nature (other than from urinary origin).

It has been suggested that the site of production of this primer pheromone is the submaxillary salivary gland. As the development of this gland is age dependent, Hughes and Kirkwood (1980) investigated this stimulus using boars of different ages (Table 6).

TABLE 5

THE EFFECTS OF THE PRESENCE OR ABSENCE OF OLFACTORY BULBS ON PUBERTY ATTAINMENT IN THE GILT (HUGHES AND KIRKWOOD, 1980)

Treatment	Age at puberty (days)	Interval between boar introduction and puberty (days)
Bulbectomy exposed to boar	$229.9^{ab} \pm 14.6$	69.9 ± 14.6
Sham operation exposed to boar	$204.0^{ab} \pm 2.5$	44.0 ± 2.5
Non-operated exposed to boar	$207.6^{bd} \pm 8.5$	47.6 ± 8.5
Control not exposed to boar	$234.4^{cd} \pm 7.8$	74.4 ± 7.8

a.b. etc. means followed by the same letter differ at $P < 0.05$

TABLE 6

THE EFFECT OF EXPOSURE TO BOARS OF DIFFERENT AGES ON PUBERTY ATTAINMENT AND OVULATION RATE IN THE GILT (HUGHES AND KIRKWOOD, 1980)

	Mean age at puberty (days)	Mean time to puberty (days)	Ovulation rate
Exposure to 2 year old boar	$182.0^{ab} \pm 5.4$	18.8 ± 5.0	13.2 ± 0.7
Exposure to 11 month old boar	$181.6^{cd} \pm 2.8$	18.3 ± 2.3	13.0 ± 0.4
Exposure to 6.5 month old boar	$206.0^{ac} \pm 1.9$	42.4 ± 1.1	12.4 ± 1.1
Controls - no boar exposure	$203.0^{bd} \pm 5.1$	39.4 ± 4.7	12.3 ± 0.6

a.b.c.d., followed by the same letter differ at $P < 0.01$

These results indicate that boar age is important in giving sufficient stimulus to induce heat in young gilts.

3.2. Sows

Anoestrus is a more serious problem in sows than in gilts. At present, the facts are:

- that the frequency of anoestrus is much higher in first-litter sows than in older sows
- that the frequency is much higher in the summer period than the rest of the year
- that a lactation period of less than three weeks delays oestrus in sows.

Although the problem is more serious with regard to sows, comparatively less research has been done on sows than on gilts. The situation with regard to housing systems, light, boar effect etc. is not clear. Modern, labour-saving housing systems in particular might be responsible for this problem. Results are contradictory. Salehar (1965), MacLean (1969) and Fahmy and Dufour (1976) found that individual housing after weaning gave the best results, whilst Tuinte (1971), Schlegel and Sklenar (1972) and Bisperink (1979) found that the group system gave the shortest interval between weaning and mating. Most of these investigations were made by collecting data on working pig farms, so that it is possible that other factors such as feed, feeding level, light, breed, management, etc. are responsible for these discrepancies. The following experiment was designed to allow a deeper study of the effect of housing system and the boar on the onset of oestrus.

4. MATERIAL AND METHODS

The experiment was divided into two studies. In both studies, four commercial farms of similar design, genetic stock (Hypor F1 sows) feeding and management practices were selected.

4.1. Study 1

There were 100 breeding sows and 4 breeding boars at each of the 4 farms. At each farm weaned sows were randomly allocated within - parity to the following post-weaning social treatments:

- I — Individual pen (60 x 192 cm). Restricted physical contact between neighbouring sows was possible through horizontal bars;
- I+(♂) — as I, but a mature boar was introduced daily into the corridor in front of the sow's pen for 10 min. Restricted physical contact between the sow and the boar was possible over the 85 cm high pen front;
- G — Group pen measuring 267 x 298 cm slatted floor, housing 5 sows;
- G+(♂) — as G, but a mature boar was introduced daily into the corridor in front of the pen, as in treatment I+(♂).

4.2. Study 2

At each of the 4 farms there were 235 breeding sows and 6 breeding boars. Sows were randomly allocated within - parity to the following social treatments at weaning:

- I — Individual pen (65 x 198 cm);
- Ig — Individual pen as in treatment I, but at weaning, sows were grouped in fours for 30 min in a pen measuring 252 x 315 cm, partly slatted floors;
- I+♂ — as I, but each sow was introduced daily into a mature boar's pen for about 3 min.
- G — Group pen (252 x 315 cm) with a partly slatted floor in the manure corridor. The manure corridor was partly divided from the laying area. Group size = 4 sows;

G+♂ as G, but each sow was introduced daily to a boar as in treatment I+♂.

Thus in both studies 2 main effects were studied:

1. housing system - individual or group
2. boar stimulation - absent or present.

In both studies the weaned sows were checked daily for oestrus by the back pressure test (Signoret, 1970) and vulva examination (Willemse, 1976). Sows in oestrus were taken randomly to a boar. At 14 and 28 days post-weaning sows that had been mated were removed from their treatments and placed in individual pens in the pregnancy section. Sows which had not mated within 28 days received an injection of a combination of PMS and HCG ('PG 600'). These sows were recorded as mated on day 29. After injection they were inspected daily to discover their response to the injection.

5. RESULTS

The reproduction performance of the sows in Study 1 is presented in Table 7.

TABLE 7

WEANING TO RE-MATING INTERVAL FOR FIRST OR GREATER-LITTER SOWS IN STUDY 1 (HEMSWORTH ET AL., 1980)

	Ind.	Ind. (+ boar)	Group (5)	Group (5) (+ boar)
1st litter sows				
Number of sows	25	26	22	25
Average interval (days)	20.96	14.04	17.59	15.48
% sows > 28 days	52.0	19.2	22.7	20.0
Older sows				
Number of sows	102	108	111	107
Average interval (days)	10.8	12.06	11.23	10.11
% sows > 28 days	10.8	15.7	8.1	8.4

P value for housing system < 0.06; for boar contact > 0.10

The main results of study 2 are presented in Table 8.

TABLE 8

WEANING TO RE-MATING INTERVAL FOR FIRST- OR GREATER-LITTER SOWS IN STUDY 2 (HEMSWORTH ET AL., 1980)

	Ind.	Ind. 30 min in group	Ind. + boar 3 min	Group (4)	Group (4) + boar (ind.) 3 min
1st litter sows					
Number of sows	31	29	29	30	27
Average interval (days)	23.90	25.28	20.76	21.97	16.37
% sows > 28 days	64.5	69.0	37.9	40.0	22.2
Older sows					
Number of sows	93	91	87	86	88
Average interval (days)	14.01	12.95	12.23	13.47	10.25
% sows > 28 days	22.6	15.4	17.2	19.8	8.0

P value for housing system < 0.05; for boar contact < 0.005

6. DISCUSSION

Comparing the results of both studies we can say that the group system is to be preferred to keeping the sows in individual boxes. The possibility of better social contact seems to be a better stimulus than the crate system. The differences between both studies for the group system may be influenced by the group size (5 vs 4 sows) or the stocking density respectively 1.6 m^2 and 2.0 m^2/sow. In addition the group pens in the second study had a partial pen division, dividing the pen into 2 areas, which may have provided an escape area for subordinate sows.

The difference in boar effect varies with age (first-litter sows vs older sows) but it is also possible that the more intensive boar contact in Study 2 is a greater stimulus for the sow than the limited contact in Study 1.

Comparing the results of the gilts and the sows it seems as if the priority of the different stimuli is not the same:

1) For gilts the odour of the boar is probably more important than physical contact with the boar.

2) For sows a limited daily physical contact is more likely to induce oestrus than is the pheromone produced by the boar.

7. CONCLUSIONS

A general conclusion can be that there is a lot of evidence in support of the idea that the female pig needs stimuli for normal reproduction. Modern piggery designs and management aspects may be responsible for both the quality and quantity of the stimuli being below a certain level for a large number of females.

REFERENCES

Bisperink, H.J., 1979. Productiviteit en rentabiliteit van zeugen LEI publ. no. 3.90. The Hague, Netherlands.

Bourn, P., Kinsey, R., Carlson, R. and Zimmerman, D.R., 1975. Puberty in gilts as influenced by boar exposure and 'transport phenomenon'. J. Anim. Sci., 41, 344 (abstr.).

Brooks, P.H. and Cole, D.J.A., 1978. In: Early sexual maturity and mating in gilts. ADAS Quarterly Review 30, 139-152.

Corteel, J.M., Signoret, J.P. and du Mesnil du Buisson, F., 1964. Variations saisonniers de la reproduction de la truie et facteurs favorisant l'anoestrus temporale. Proc. 5th Int. Congr. Anim. Reprod. Art. Ins. 536-540.

Fahmy, M.H. and Dufour, J.J., 1976. Effect of post-weaning stress and feeding management on return to oestrus and reproductive traits during early pregnancy in sows. Anim. Prod. 23, 103-110.

Hemsworth, P.H., Hoogerbrugge, A. and Salden, N.T.C.J., 1981. The post-weaning social environment and reproduction in the sow (in press).

Hughes, P.E. and Kirkwood, R.N., 1980. Boar-induced precocious puberty in the gilt. Proc. of the 31st European Congress of EAAP no. P 5/6.39 Munich.

Hurtgen, J.P., Le Man, A.D. and Crabo, B., 1980. Seasonal influence on oestrus activity in sows and gilts. JAVMA 176: 2, 119-123.

Karlberg, K., 1980. Factors affecting postweaning oestrus in the sow. Nord. Vet. Med. 32, 185-193.

King, G.J., 1978. Occurrence of postweaning estrus in primiparous and pluriparous cows. Proc., 5th World Int. Pig Vet. Soc. Congr. Zagreb, Paper KA 12.

Kirkwood, R.N. and Hughes, P.E., 1980. A note on the efficacy of continuous v. limited boar exposure on puberty attainment in the gilt. Anim. Prod. 31, 205-207.

Love, R.J., 1979. Reproductive performance of first parity sows. Vet. Rec. 104, 238-240.

MacLean, C.W., 1969. Observations of non-infectious infertility in sows. Vet. Rec. 85, 675-682.

du Mesnil du Buisson, F, and Signoret, J.P., 1962. Influences de facteurs externes sur le déclenchement de la puberté chez la truie. Ann. Zootechn. 11, 53-59.

Rasbeck, N.O., 1969. A review of the causes of reproductive failure in swine. The Brit. Vet. Journ.

Salehar, A. Conception in sows after weaning the litters (cited by Falmy and Dufour).

Schlegel, W. and Sklenar, V., 1972. Der Einfluss unterschiedlichen Haltungsformen auf die Fortpflanzungsleistung von Sauen. Tierzucht 26, 409-410.

Signoret, J.P., 1975. In: The Behaviour of domestic animals. Ed. Hafez Bailliere Tindall, London, pp 295-329.

Stork, M.G., 1979. Seasonal reproductive inefficiency in large pig breeding units in Britain. Vet. Rec. 104, 49-52.

Tuinte, J.H.G., 1971. Uitval van zeugen bij verschillend huisvestingsmethoden Maandblad van de Varkensfokkerij, 33, 201-203.

Tuinte, J.H.G., 1976. Wel of geen uitloop voor fokzeugen. Proefverslag no. 10. Raalte.

Willemse, A.H. and Boender, H., 1976. Quantitative and qualitative analysis of oestrus in gilts. Tijdschr. Diergenessk. 91, 349-354.

Zimmerman, D.R., Bourne, P. and Donovan, D., 1976. Influence of boar component stimuli on age of puberty in gilts. J. Anim. Sci. 42, 1362 (abstr.)

DISCUSSION

J.P. Signoret *(France)*

Did you make some hormone assays during the period directly after contact, or different type of contact, with the boar? Have you some estimations of the ovarian or genital tract development of the females that you are submitting to boar contact?

A. Hoogerbrugge *(Netherlands)*

No. This year we are assaying blood progesterone, but that is only at the eighteenth day after weaning to examine sows that did not come on heat to see if they have been in a silent heat or not, then on the 21st day we inject the anoestrus females using gonadotrophic hormones to see if they are coming into heat or not. It is a kind of diagnostic injection and also gives information about whether there has been a silent heat.

SOME ASPECTS OF SOW HEALTH AND HOUSING

I. Ekesbo

Department of Animal Hygiene, Faculty of Veterinary Medicine,
Swedish University of Agricultural Sciences,
SKARA, Sweden.

1. INTRODUCTION

The aim of this paper is not only to throw light on some aspects of sow health and housing but also to force ourselves to think about and try to answer the following three questions:

1) Has the development of swine environment - housing and management - been beneficial in all respects from the point of view of animal health and welfare during the last two decades? In other words, are we using the possibilities of modern technology in an appropriate way?

2) Is the increased use of antibiotics or other drugs the best method of fighting diseases partly or totally caused by environmental factors?

3) If the answer to these two questions is 'no', what can we do to change the situation?

2. HEALTH AND HOUSING

In discussing sow health and housing, it might be interesting to make a comparison of the situation as it was 20 years ago, 10 years ago and today. I will first give a short account of the sow environment in Sweden, the development of which might be typical of that in most countries from the end of the 1950s to the beginning of the 1970s and so give us some sort of perspective on the situation.

However, to avoid misunderstanding I would like to draw attention to the fact that, as a rule, environmentally evoked diseases are caused not by a single factor but by a combination of factors (Figure 1).

CLINICALLY MANIFESTED HERD DISEASE	*E.g. abscesses in sows and/or piglets*
SUBCLINICAL HERD DISEASE CONDITION	LARGE HERD OR GROUP SIZE SLATTED FLOOR IN SOW-PIGLET PEN ROUGH CONCRETE FLOOR SURFACE UNSUITABLE PEN DEVICES SOWS KEPT CONFINED NO STRAW INFECTION
HEALTH	

Fig. 1. Aetiology of partly or totally environmentally evoked herd disease problems. Fundamental elucidation.

The type of animal husbandry which was traditionally practised until around 1960 was the result of the accumulated knowledge of generations of farmers of what is practicably applicable in the keeping of animals. Twenty years ago there was only traditional sow housing.

Traditionally, dry sows were kept in individual pens or, in large herds, in pens for small groups. This system made it possible for the sows to use separate parts of the available pen area for the three functions of bedding, dunging and feeding. The environment was thus, in this aspect, adapted to basic biological needs of the swine.

The traditional pen for a sow with a litter also fulfilled this requirement of free choice. Although the buildings and the fittings were often quite simple, the animals' behaviour made it possible for them to be kept clean as a result of free access to a separate dunging area.

2.1 Housing requirements

Man seems to have understood from the beginning that the pig requires separate feeding, dunging and lying areas. It appears from the literature on the subject (Brauner, 1772) that man never used a system whereby the pig was tied or permanently confined in such a way that it could not separate these three functions. This cannot be a coincidence.

Nervous sows, or sows which for other reasons could hurt the piglets, were confined to cages of different type for a few days. Although this facility was always available it was more an exception than a rule in most herds.

Feeding, watering and manure handling were all manual labour. However, about 90% of the sows usually dunged in the same part of the pen which reduced the labour.

It is obvious that in traditional animal husbandry, a result of the accumulated knowledge of many generations of farmers, great attention was paid to the cleanliness of pigs. In the rather abundant 18th century literature on agriculture (Brauner, 1772) the importance of providing the swine with a dry lying place with straw bedding is continually stressed. There is even a description of how the problem of straw bedding should be solved during periods when straw was scarce. Obviously, the presence of straw in the lying area was considered necessary for the well-being and health of the animals.

2.2 Technical demands

The beginning of the 1960s more or less brought about a revolution in sow environment (Ekesbo, 1973). Design of the animal environment was no longer a product of co-operation between farmer and sow, it was often created at the drawing-board where technical demands were given a higher priority than the animals' basic needs.

One consequence of this revolution was that sows in many herds no longer had free choice within the area available for bedding, dunging and feeding. The confinement system forced sows to dung in their bedding area. Several different confinement systems came into use, e.g. the Protecta system, the neck-tying system, the girdle-tying system and the crate system with or without slatted floor. For those of us who were in close contact with numerous commercial farms during these years, one thing was very obvious: the sows in the new confined systems were much dirtier than the sows in the previous system.

During these years the subjective observation was made that when sows had a free choice of keeping dunging, feeding and bedding areas separated from each other, less than 10% of the sows were dirty. When, however, the sows were kept confined more than 75% of them were more or less dirty. This is also the case when a slatted floor was used, even though the degree of dirtiness might be less than without a slatted floor. These facts are not changed when a solution is sought in the use of other floor systems.

If, however, the sows' standard of hygiene was lowered, development during the 1960s no doubt produced many animal houses with a more uniform and clean appearance than some of the old houses with their home-made fittings. However, the greater number of animals/area increased the need for ventilation which created noise problems (Algers et al., 1978a; Algers et al., 1978b) and sometimes problems of draught and low relative humidity. Some of these changes combined to create a dust problem as well. Where liquid manure handling was introduced the animals were often exposed to toxic gases and straw was no longer used.

2.3 Distress factors

To sum up: the animals were exposed to what might be called distress factors -

a) when their normal behaviour pattern of feeding, dunging and bedding in separate areas was prohibited,

b) when their environment became more barren, with a risk of understimulation, and

c) when they were exposed to overstimulation, e.g. through constant noise.

2.4 Environment and health

During the second part of the 1960s Bäckström (1973) carried out an investigation of the influence of the environment on swine health comprising about 9 600 sows and about 9 100 litters from 834 herds. He found a definite connection between herd size and increased frequency of *agalactia toxaemia*, traumatic injuries in sows, piglet morbidity and piglet mortality. Bäckström explains these facts in two ways. Firstly, the herd size itself can be a factor in the spreading of infection. Secondly, however, the relationship between morbidity and herd size can also, in certain respects, reflect a combination of different environmental factors as well as hygiene and management factors.

Bäckström's studies of environment factors showed that the health of litters in pens with straw as bedding was superior to that of litters in pens with other bedding materials. Litters in pens without bedding showed the poorest results. Herds in houses with liquid manure handling systems were associated with poorer pig health than those with solid manure handling systems. Bäckström points out that these results generally correspond to those reported in the literature on fattening pig production. Bäckström explains this partly as a toxic effect of manure gases and partly as an effect of lack of bedding in herds with liquid manure handling.

Sows in pens with slatted floors showed a higher incidence of foot lesions than sows in pens without slatted floor (Table 1). Also, piglet morbidity and mortality were higher in pens with slatted floors than in pens without (Table 2). Systems of husbandry with maximum restraint of sows during pregnancy had more problems with sow illness at farrowing than herds where the sows had maximum freedom of movement (Table 3). Increased frequency of piglet mortality which could be related to foetal development and farrowing illness of the sow also occurred when the animals were restrained (Table 4).

Systems where pregnant sows had some form of outdoor exercise showed results that were superior to those where the sows were kept free in pens indoors during gestation. The sows reacted unfavourably to any change of environment at farrowing, irrespective of its nature.

TABLE 1

NUMBER OF SOWS DURING GESTATION AND LACTATION AND INCIDENCE OF SOW FOOT LESIONS IN PENS WITH OR WITHOUT PARTLY SLATTED FLOORS

	The floor in pens or crates		Statistical difference
	without slats	partly slatted	
Number of sows	3 520	588	***
% foot lesions	3.3	6.3	

(Bäckström, 1973)

TABLE 2

NUMBER OF LITTERS, NUMBER OF PIGLETS BORN, PIGLET MORTALITY AND LITTER MORBIDITY IN PENS WITH CRATES WITH OR WITHOUT PARTLY SLATTED FLOOR.

	Pen floor		Statistical difference
	without slats	partly slatted	
Number of litters	1 393	681	
Number of piglets born	15 902	7 906	
% total mortality	9.0	11.9	***
% diarrhoea	10.8	17.6	***
% 'unthriftiness'	18.3	26.0	***
% total morbidity	57.2	62.8	*

(Bäckström, 1973)

TABLE 3

NUMBER OF SOWS AND SOW MORBIDITY AT FARROWING IN RELATION TO THE DEGREE OF SOW CONFINEMENT DURING THE GESTATION PERIOD AND AT FARROWING.

	Sow free in pen	Sow confined in crate	Statistical difference
Number of sows	654	1 283	
% *agalactia toxaemica*	6.7	11.2	**
% farrowing time > 8 h	2.3	5.4	**
% total morbidity	12.8	24.1	***
% medically treated	10.4	16.8	***

agalactia toxaemica = mastitis metritis agalactia (Bäckström, 1973)

TABLE 4

PIGLET MORTALITY IN RELATION TO SOW CONFINEMENT DURING THE GESTATION PERIOD AND AT FARROWING.

	Sow free in pen	Sow confined in crate	Statistical difference
Numbers of litters	554	1 079	
Number of piglets born	6 307	12 534	
% stillborn	5.1	6.3	***
% mummified	0.1	0.7	***
% splaylegs	3.1	3.9	**
% result of ailing sow	2.4	3.4	***

(Bäckström, 1973)

2.5 Morbidity and mortality

Piglet morbidity and mortality were shown to be affected by both floor area and pen fittings during the lactation period. The lowest morbidity was clearly associated with litters in pens of > 6 m^2 floor area. Piglet mortality was lower and the number of piglets surviving/litter higher with larger floor areas. The results became successively poorer with smaller floor area. The differences were more marked in the older litters. Poorer health during the first three weeks was shown in litters where the sows were permanently crated than in litters where the sows were free in pens. Concerning total piglet morbidity, particularly great differences occurred between large and small pens (> 5 respectively < 5 m^2) where the sows were restrained at farrowing and for a short time thereafter. Litters from untethered sows in large pens showed a significantly lower morbidity than litters from tethered sows in small pens.

2.6 Traumatic symptoms

The incidence of traumatic injuries in tied sows was high. Even for crated sows there was a relatively high incidence of pressure injuries compared to that in sows in pens (Table 5). Spilt water and residual dung and urine were the most common causes of sores occurring over the spine of the scapula in thin sows. However, crated sows also showed other traumatic injuries.

TABLE 5

INCIDENCE OF TRAUMATIC INJURIES CAUSED BY PEN FITTINGS AND FLOORING IN RELATION TO NUMBER OF SOWS FREE IN PENS AND CONFINED IN CRATES RESPECTIVELY.

	Sow free in pen	Sow confined in crate	Statistical difference
Number of sows	2 464	1 678	
Traumatic injuries	0.8	6.1	***

(Bäckström, 1973)

2.7 Developments

Thus, in many respects, the changes in swine environment in Sweden during the 1960s had detrimental effects on the health and hygiene of the animals. During the 1970s this development has to some extent changed. In the last few years in particular many farmers have tried to abandon the different restraint systems for the animals when rebuilding or building new piggeries and also in other respects endeavoured to meet the animals' elementary needs for the maintenance of a good state of health. Some other examples of these changes are the use of straw, the use of scrapers instead of liquid manure handling, the reduction of noise from ventilators and giving the sows and piglets free choice of lying, dunging and feeding areas (Ekesbo et al., 1978). The reason for this development might partly be the farmers' own experiences and partly advice based on reports concerning the impact of environment on animal health (Algers et al., 1978a; Algers et al., 1978b; Bäckström, 1973; Ekesbo, 1975; Högsved, 1969; Lindquist, 1974). For dry sows, as a rule, free choice is given between lying and dunging area, sometimes also between lying, dunging and feeding area.

Studies recently carried out in Sweden (Ekesbo and Holmstedt, in press) show that sows confined in cages have a higher degree of dirtiness than sows free in pens (Table 6).

TABLE 6

MEAN DEGREE OF CLEANLINESS/SOW IN HERDS WITH DRY SOWS IN CAGES, IN STALLS WITH DUNGING ALLEY AND FREE IN PENS WITH OR WITHOUT ACCESS TO OUTSIDE YARD. DEGREES OF CLEANLINESS: 1 = COMPLETELY DIRTY ANIMALS, 5 = COMPLETELY CLEAN ANIMALS.

Housing environment	Number of herds	Number of animals	Degree of cleanliness
Cage	11	1 019	3.1
Stall + dunging alley[1]	11	650	4.3
Free in pens only indoors	21	1 028	4.7
Free in pens and outside	16	947	4.7

[1] Of these herds 5 with 335 sows consisted of breeding herds. The mean 'degree of cleanliness' for these 5 was 4.7 and for the remaining 6 3.7.

It is possible that certain clinical illnesses may, to a great extent, have their origins in the fact that the animals cannot satisfy elementary behaviour needs. If so, this can be expected to show primarily in different behaviour deviations caused by frustration and distress. One example is stereotyped movement patterns in sows in barren environments, e.g. bar biting (Ekesbo, 1976; Ekesbo, 1978; Jensen, 1979; Sambraus et al., 1978). Similar symptoms are shown by early weaned piglets in flat decks (Algers, 1980).

2.8 Anoestrus

England and Spurr (1969) found that keeping sows in cages might considerably influence the onset of heat in gilts. Of gilts in cages 17% did not show heat, while the corresponding figure for gilts free in pens was 6%. The frequency of irregular oestrus was also higher for gilts in cages, 28%, than for sows free in pens, 16%. No difference was found between sows kept free in pens and sows in cages in showing oestrus, but while all covered sows kept free in pens farrowed, 7% of the sows in cages never farrowed.

2.9 Activity

In recent investigations (Ekesbo et al., 1978; Ekesbo et al., 1978; Jensen, 1979; Jensen, 1980a; Jensen, 1980b; Jensen, in press) comparing the behaviour of sows kept in crates and sows kept loose during the whole dry period, some interesting results have been noted.

The results show that tied sows are significantly less active during daytime than loose sows. At 68% of the daytime observations the tied animals were lying down while the corresponding figure for loose animals was 41%.

2.10 Biased behaviour

During the animals' active time certain differences in behaviour were registered. The tied sows showed, for example, a higher frequency of 'chewing on iron bars', 'eating/drinking' and 'sitting'. The behaviour patterns 'rubbing oneself against the fittings' and 'scratching in straw' were significantly more common in loose sows. The tied animals immediately and intensively ate all the straw while the loose sows only ate part of it and at the same time carried out 12 different behaviour patterns involving the straw. In the tied sows only two, seldom three, behaviour patterns with straw handling were shown.

It is in the nature of the restrained situation that the animals are forced to use the same space for their whole behaviour repertoire. The loose animals, however, divided their time in the three available areas; lying area, dunging area and feeding area. It should be pointed out that the loose animals were locked into the individual feeding stalls for about 3 h in the morning for feeding and cleaning, i.e. 13% of the 24 h. Loose sows, able to choose their whereabouts relatively freely, spent only 25% of the time in the feeding stalls and only 3% in the dunging area. There is reason to believe that the animals are highly motivated in executing different behaviour functions in different areas. For this reason we need systems which do not confine the sows as tying and crating do.

With respect to social behavioural patterns in the 20 loose sows, 1 216 social interactions were noticed (12 different ones) and in the 16 restrained sows 236. In the restrained animals three behaviour patterns were completely missing while three types of interaction mainly corresponded to most of the registered behaviour patterns.

The behavioural investigations showed that restraint changed the quantitative scope of certain general as well as social behaviour patterns, brought about a qualitative change of certain social behaviours, made it impossible for the animals to execute their actions in different areas, for which they probably are highly motivated. Finally, the study showed that restraint brought about a greatly increased frequency of stereotyped behaviour which can be interpreted as a sign of general frustration caused by the three situations described above.

3. DISCUSSION

It can be concluded that sows kept in confinement are exposed to more stress factors than sows kept free in pens. If other environmental factors are also detrimental, the risk of a higher incidence of not only subclinical but also clinical ailments in restrained sows must not be underestimated. Some of the findings made by Bäckström might thus to some extent be explained in this way.

At the same time as the system of confining sows, whether tied or in cages, has become more and more frequent during the 1960s and the 1970s, an increase in the frequency of disease in sows has been noted. Different ways of solving this problem have been recommended, e.g. treatment of all animals with antibiotics (Bogner et al., 1978). It is thus not unusual in Europe that the keeping of sows in cages is combined with 100% medication to prevent disease in connection with farrowing.

Thus, we have also given a negative answer to the two questions a) and b) at the beginning of this paper. To the third question c) "what can we do", I would like to propose the following. In terms of national economy as well as of ethics it is important from the point of view of animal health that optimal use is made of technical aids in the animal environment. This cannot generally be said to be the case in today's swine husbandry. We need more research into the impact of animal environment on animal health and behaviour. There is today a risk that wrong advice may cause the farmer great losses. He might, for example, be advised to omit straw in order to save money. The reason for such advice would be that the cost of straw is possible to calculate while it is impossible, or at least difficult, to calculate losses due to unhealthy conditions caused by keeping the pigs without straw. The farmer must be given a better biological base for his decisions than the one he has today.

In order to reach this goal, it is absolutely necessary to achieve a better co-operation between veterinarians, farmers and technicians so that the animals' elementary requirements for good health are met, when technical aids for use in the animal environment are designed. In Sweden, such co-operation has been statutory since 1973 in that plans for all new and reconstructed buildings must be examined and approved by specially trained veterinarians before such construction may be undertaken. In connection with this scrutiny it is also statutory, if judged necessary from the point of view of animal health and welfare, that new methods, systems and/or fittings be tested before they are introduced into common practice. The Swedish parliament has also established a special scientific section for such testing within the department of animal hygiene at the veterinary faculty.

4. CONCLUDING REMARKS

When evaluating methods, systems and/or fittings from the point of view of animal welfare, different parameters are used according to what is to be investigated, e.g. systematic clinical observations, systematic behavioural studies, epidemiological studies, autopsies, systematic studies of slaughtered animals etc. This scrutiny, and when necessary testing, excludes a certain number of factors which cause disease from the animal environment. Thus the animals are protected from these diseases caused by the environment and the farmers from losses through such diseases. As these regulations do not stop or prohibit technical development but only stipulate that it fulfils the biological requirements of the animals, they constitute a protection both for the animals and for the farmers.

absense of stress detrimental to health

REFERENCES

Algers, B., 1980. Animal health and behaviour in flatdeck rearing of weaned piglets. Swedish Univ. of Agr. Sc., Faculty of Vet. Med., Dep. of animal hygiene, Report 4, Skara.

Algers, B., Ekesbo, I. and Strömberg, S., 1978a. Acta Vet. Scand., Suppl. 67.

Algers, B., Ekesbo, I. and Strömberg, S., 1978b. Acta Vet. Scand., Suppl. 68.

Bogner, H., Schumm, H., Berner, H. and Matzke, P., 1978. Tierärztl. Umschau, 33.

Brauner, J., 1772. Skötseln och nyttan af boskap och fjäderfä, Stockholm.

Bäckström, L., 1973. Acta Vet. Scand., Suppl., 41.

Ekesbo, I., 1973. Vet. Rec., 93.

Ekesbo, I., 1975. Miljö-djurhälsa. Kompendium Sveriges Veterinärförbund, Allmänt Veterinärmöte.

Ekesbo, I., 1976. Etik och etologi i animalieproduktionen. Skogs- och Lantbruksakademiens Tidskrift, 115.

Ekesbo, I., 1978. Proceeding 1st World Congress on Ethology Applied to Zootechnics, Madrid.

Ekesbo, I. and Holmstedt, K. (in press).

Ekesbo, I., Jensen, P. and Högsved, O., 1978. Svensk Veterinär Tidning, 30.

Ekesbo, I., Jensen, P. and Lock, R., 1978. Svensk Veterinär Tidning, 31.

England, D.C. and Spurr, D.T., 1969. J. Anim. Sci., 28: 2.

Högsved, O., 1969. NJF Seminar Kalnes Sarpsborg (Norge), September 1969.

Jensen, P., 1979. Behaviour patterns of dry sows in three different housing systems - a pilot study. Sw. Univ. of Agr. Sci., Dept. of Animal Hygiene, Report 1, Skara.

Jensen, P., 1980a. The effect of confinement on the behaviour of dry sows - an ethological study. Sw. Univ. of Agr. Sci., Dept. of Animal Hygiene, Report 2, Skara.

Jensen, P., 1980b. An ethogram of social interactions in grouphoused dry sows. Applied Animal Ethology, 6.

Jensen, P., (in press).

Lindquist, J.-O., 1974. Acta Vet. Scand., Suppl., 51.

van Putten, G., 1969. Br. Vet. J., 125.

Sambraus, H.H., Sommer, B. and Kräusslich, H., 1978. Proc. 1st World Congress on Ethology Applied to Zootechnics, Madrid.

DISCUSSION

G. van Putten *(Netherlands)*

You stressed the point that we need to do more research in this special field. Is it your opinion that we ought to wait in regard to regulations until we have completed more research?

I. Ekesbo *(Sweden)*

My final paragraph expresses my opinion on this. In Sweden we followed the consequences of the situation as early as 1973. This was due to a group of veterinarians who felt a great responsibility that new frontiers must be opened to fight environmentally evoked diseases. The way to do this is by more co-operation with technicians, facts on the table, but the farmer should not have to pay for what should be financed by others.

A. Hoogerbrugge *(Netherlands)*

Some of our investigations have revealed aspects that are not in favour of the grouping of sows. First, there was more fighting amongst the sows, and secondly there was more leg weakness.

W.T. Truyen *(Netherlands)*

In the Netherlands we have used the system on many farms where there is a small lying area, just enough for 4 or 5 sows, and a marked dunging place with a slatted floor. Most of the farmers have changed to other systems because of the habit of the sows to lie down on the slatted floors rather than their lying areas, and then dunging in the sleeping area. Does this occur in Sweden?

I. Ekesbo

We have two systems. One has a very well marked dunging area and a very small laying area of about 1 m^2/animal.

The other system has very simple uninsulated buildings where there is a large bedding area which is cleaned every second or third month with a tractor. The farmer leaves the animals to choose their own dunging place, and just gives the straw. In the last five or six years the tendency is definitely towards this system.

SESSION VI

COMPARATIVE DATA REGARDING WELFARE IN PRESENT HUSBANDRY SYSTEMS, II.

2. SUCKLING SOWS WITH PIGLETS

Chairman: D. Smidt

THE IMPACT OF THE FARROWING PEN ON THE SOW AND HER PIGLETS

J. Daelemans
National Institute for Agricultural Engineering,
Merelbeke, Belgium.

1. INTRODUCTION

During the last two decades the housing and fittings for pigs have changed steadily. The development of the farrowing pen with straw from the transferable sow crate in 1960 started a remarkable pattern of change. To start with, sows were fed in stalls outside the farrowing pen. Providing the farrowing pen with a sow trough was the first step towards efficient housing. Since 1970, further steps have been made: reduction of pen dimensions as a consequence of earlier weaning, the introduction of partly and even fully slatted floors, the elimination of the use of straw, and the restraint of the sow, together with the installation of turning crate separations. Nowadays the development of wire and plastic slats has opened possibilities for more acceptable housing conditions for pigs, whilst at the same time compartmentation and batch handling of herds lead to opportunities for better organisation of the labour force. This continuous evolution, offering a variety of housing systems, and the growing influence of animal welfare research complicate decision-making for pig farmers. This paper looks at some aspects of pen flooring and pen shape.

2. REQUIREMENTS

The purpose of a farrowing pen is to house a sow in the last days before farrowing, during farrowing, and even after farrowing, when the pen will contain the litter at least until weaning, and sometimes even after that time. By this time, requirements with respect to building materials, equipment and environment are difficult to meet. The sow requires a different temperature to that demanded by piglets, whose needs change in relation to their age. The same must be stressed with regard to somatic characteristics. Baxter (1971) has already related the slot size of a slatted floor to pig size. Finally, choice of materials is important in connection with the prevention of heat loss and injuries to young animals which are not as well protected as older ones.

3. FLOORING IN FARROWING PENS

Maton et al. (1979) reported a comparative study of claw and carpal lesions in suckling piglets in farrowing pens with partly slatted concrete floors. The alternatives compared were: 1) a concrete floor with sawdust, 2) a concrete floor with straw and 3) a rubber mat. All piglets were individually examined once a week for six weeks, and the following conclusions were drawn:

1) Approximately 92% of the piglets showed no claw lesions at all. There was no significant difference between the three floors (92.5%, 92.3% and 91.3%). In 7 - 8% of the piglets there was a slight erosion of the toe-points, but this was also found to be insignificant.

2) Serious claw and carpal lesions were rare, and during the entire observation period were not significantly more frequent on one floor than on another.

3) In the 1st and 2nd weeks after birth claw lesions can regularly be observed in piglets, but rarely during the 3rd and 4th weeks, irrespective of the flooring on which the piglets are kept.

4) Piglets often suffer from light lesions on the carpus in the 1st month of their lives, irrespective of the type of floor on which they are kept; these injuries are more frequent and last longer than claw injuries.

5) From a practical point of a view, housing piglets on concrete floors slightly strewn with sawdust presents no disadvantages with respect to claw and carpal lesions; this method of housing saves both on straw and labour in comparison with littered pens and on materials in comparison with rubber mat flooring.

Similar conclusions have been drawn by Gravås, who could observe no significant differences in the frequency of carpal lesions in piglets kept on an isolated concrete floor lightly covered in sawdust, on an isloated floor covered with a 5 mm rubber mat and on an epoxy-resin treated isolated floor. Piglets kept on a rubber mat, however, showed deeper and larger carpal lesions than those which were kept on a concrete floor.

In view of the recent changes in floor and slat materials which are available (wire and plastic, plastic-coated concrete or metal), similar research would give more information about the equipment which is available on the market today.

4. TYING THE SOW

Koomans (1967) compared dry and pregnant sows housed in tied stalls with loose sows kept in a group and found similar production results. Tuinte (1976) found slightly different results, with the better results coming from the group with an outside exercise area. However, when the dry sows were kept loose in an indoor group pen and afterwards in individual stalls, the results were comparable with those of the loose sows which had access to an outside exercise area.

Jeppsson et al. (1980) described the behaviour of loose non-suckling sows and non-suckling sows shut up in stalls under the same housing conditions. They showed that loose dry and pregnant sows spent 80% of a 24 h day lying, whilst fixated sows spent only 76% lying and that this difference was significant. Loose sows spent significantly more time lying on their sides than fixated ones. The authors concluded that "of the two housing systems for dry and pregnant sows studied, the loose housing system was most acceptable for the animals from the view-point of behaviour because sows maintained in that system had the best opportunities to keep themselves clean, they could choose their location and lying position, they were not forced to dung or urinate on the eating area, and the loose sows also had a certain access to 'exercise' in comparison to the fixed sows".

This important statement in favour of loose housing for sows is not applicable in the case of suckling sows. Indeed, where the sow has to deliver and grow her piglets in a farrowing pen, other requirements predominate. In these circumstances the sow's movements have to be thoroughly restricted. Only small sow crates or tied up stalls, correspondingly enlarged with the necessary piglet area, are adequate farrowing devices and protect the piglets as much as possible from being crushed.

Pflug (1976) studied farrowing sows and their piglets during the first three days of life. He stated that there is no day-night rhythm, which is contrary to the data for dry sows as published by Jeppsson et al. (1980). Later in his study, Pflug (1976) detailed more advantages of the farrowing crate as opposed to the tied farrowing pen, although he nevertheless did not judge the latter as being inadequate for the sow. The author stresses the necessity of having a good micro-climate for the piglets in the creep "as otherwise the piglets rest near the sow and run the risk to be squeezed to death".

The restricted place of the sow in the farrowing pen allows restricted total pen dimensions if the parallel or diagonal area for the piglets is also limited (Eichhorn, 1978). Actual farrowing pens are therefore small in size, with parallel or diagonal crates, permitting heating at the lowest possible cost.

5. WATER CONSUMPTION

5.1. The water consumption of suckling piglets

Bekaert et al. (1970) compared three different ways of giving water to piglets: open water cups, automatic drinking bowls and nipples.

The piglets did not drink durng the first week. Only the piglets with the simplest water dispenser, the open water trough, consumed a certain amount of water during the second week, whilst the piglets which had one of the other two distribution systems either drank very little or did not drink at all. This can be explained by the fact that it was easiest for the piglets to drink from an open water cup. From the third week there was an appreciable amount of water consumed with all systems except the nipple system with indirect valve stem action. With this system, the piglets only began to drink during the fifth week of life. There was a large amount of wasted water with the nipple system and very little, if any, with water bowls or open water cups.

The most interesting water distribution systems for piglets would seem to be special automatic water bowls and nipple drinkers with direct valve stem action.

5.2. The water consumption of suckling sows

Suckling Belgian Landrace sows were given a ration adapted to the number and age of piglets. Later, water consumption was closely related to milk production, which increased with the age of the piglets. Measurements showed that the ratio of water to feed remained at 3.5 during suckling (3.5 l water consumed/kg feed); this was about 40% more than the ratio for pregnant sows. This resulted in a relatively high water consumption which increased over the time after farrowing until the 6th week (from 10.2 l/sow/day to 16.2 l/sow/day during the 5th week). Indeed, during the 6th week the piglets were only given access to the sow for a short period twice a day. At the same time, feed was restricted and water consumption fell to 11.2 l/sow/day.

6. CONCLUSIONS

Over the last two decades the construction of farrowing pens has changed radically, and it seems as if this trend towards renovation has not yet ended. The elimination of straw has fulfilled the goal of a more humane method of pig housing by reducing the amount of labour required and the amount of physical effort. The elimination of straw is, however, only made possible by the provision of adequate floorings and equipment and a proper climate. Future developments of better floorings may lead to housing better adapted to both sow and piglets.

REFERENCES

Baxter, S.H. and Robertson, A.M., 1971. Slatted floors in farrowing and rearing pens. Farm Buildings Progress, 26 October 1979, 27-36.

Bekaert, H. and Daelemans, J., 1970. De drinkwatervoorziening bij zogende biggen. Landbouwtijdschrift 23: 925-939.

Eichhorn, H., 1978. Haltingstechnik bei ferkelführenden Sauen und Ferkelaufzucht. KTBL-Schrift 227: 84-98.

Gravås, L., 1977. The effect of the floor on behaviour and knee damages on piglets and sows. Meldinger fra Norges Landbrukshogskole.

Jeppsson, M., Svendsen, J. and Andreasson, B., 1980. Behaviour studies of 'loose' and 'fixed' dry sows, maintained under the same husbandry, feeding and stable conditions. Institutionen för lándbrukets byggnadsteknik Lund.

Koomans, P., 1967. Vastgebonden zeugen. Mededeling 24 ILB-Wageningen.

Maton, A., Daelemans, J. and Hoorens, J., 1979. The influence of housing conditions on claw and carpal lesions in suckling piglets.

Pflug, R., 1976. Geburtsverhalten von Sauen und Verhaltensweisen ihrer Ferkel. KTBL-Schrift 208.

Tuinte, J.H.G., 1976. Wel of geen uitloop voor fokzeugen. Bedrijfsontwikkeling 7: 909-910.

DISCUSSION

P. Koomans *(Netherlands)*

How do you measure the amount of waste water, especially from the nipples?

J. Daelemans *(Belgium)*

Wastage measured with a water meter from indirect valve nipples was about 28%, but with direct valve ones the mean was 45% and the highest was 72%; so there was a large amount of wastage.

WELFARE AND THE HOUSING OF THE SOW AND SUCKLING PIGS

S.H. Baxter

Scottish Farm Buildings Investigation Unit,
Craibstone, Bucksburn, Aberdeen AB2 9TR, Scotland.

1. INTRODUCTION

Whilst it is clear that the subject matter of this paper will be dominated by a consideration of one rather narrow aspect of the present situation in pig production, the broader influential factors of the past and the apparent trends of the near future cannot readily be discounted. Indeed the future direction of pig production will be the result of a large number of pressures and responses dominated by the set of major interacting systems shown in Figure 1.

At the heart of the pig meat production enterprise there is a biological system composed of a number of sub-systems like the anatomical, physiological, hormonal and behavioural. In a broader perspective however the biological system is subject to the control of the management system. Pigmeat production is the control and manipulation of a biological system by man (management) for purposes and ends prescribed by him but attainable only within the ultimate biological capabilities of the system. Management is controlled and manipulated to a great extent by the economic system. Here, financial reward or penalty is the ultimate motivator for management to extend its control of the biological system. Finally however even the economic system is subject to the vagaries of the complex and variable political system. It is perhaps significant that it is in the political arena that concerns for animal welfare have been most forcibly expressed (Brambell, 1965)*. It would be naive to assume that animal welfare can be considered in the sole context of the interaction of management and biological systems and only focused on such zoocentric issues as cruelty and discomfort. There

* In UK see also Codes of Welfare for Pigs; and the various manifestos of the political parties prior to the General Election of 1979.

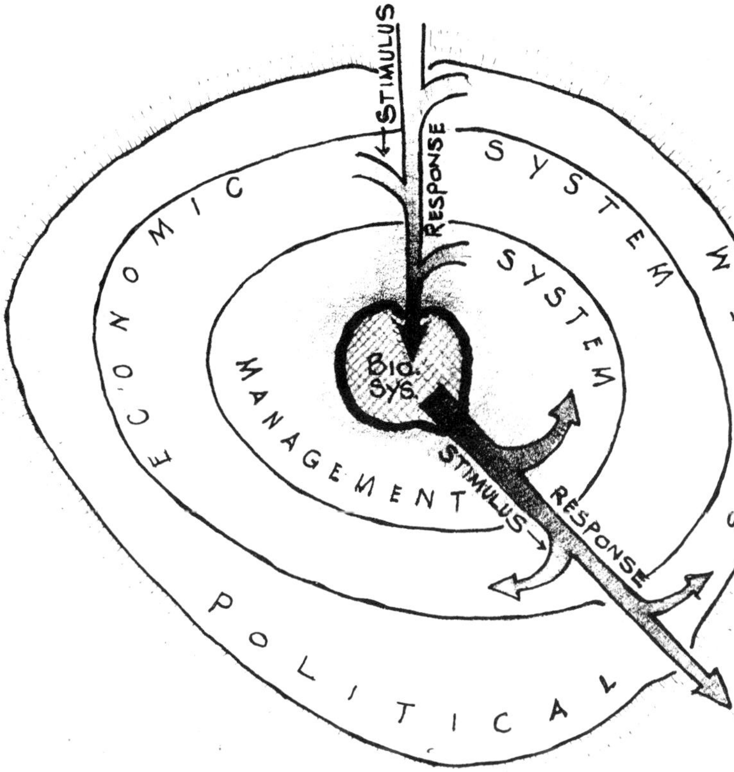

Fig. 1. Systems within Systems.

is a need for a broader philosophy of welfare encompassing both biological and moral concerns within a realistic economic model imbued with political commitment.

Faced with these unremitting external pressures yet encouraged by the inexorable advances of science and technology pig production has, almost inevitably, moved along its present path. In the last 20 years major changes have occurred in both the structure and operational aspects of pig production. The slow demise of the very small pig enterprise has been over-shadowed by the explosive growth in size of the larger units. Traditionally the notorious pig cycle was in part maintained by the tactics of the smaller producer. When profits were poor he easily moved out of pigs and when they were good he just as easily moved back again. The larger, more financially committed erprise however sought protection in both economic conditions by expansion. When profits were good there was more to be made in bigger units, when profits were poor total return was maintained by an increase in output to counter the lower profit per unit of output. A continued escalation of this trend in a period when supplies advanced faster than *per capita* consumption may have forced out even more small scale producers.

Expansion requires high levels of investment on a scale commanded by only the largest operations, and these in turn get even larger. Such investment has been directed into areas of scientific and technological advance, substituting buildings and equipment for labour and supporting the implementation of the latest scientific ideas. Productivity has been increased by the rapid and large scale adoption of improved breeding, controlled nutrition, new husbandry practices like earlier weaning, preventive medicine and environmental modification. Inevitably this concentration of technology on what is essentially a biological process has become known as intensive livestock production or even factory farming in the most extreme cases. Characterised by an increase in capital investment in housing and equipment, a reduction in labour usage, an increase in the nutrient intake per animal to achieve a high rate of output and a standardisation of the final product, (Blaxter, 1967) the intensive pig unit has developed. These units are larger and more concentrated in their use of land than the more traditional farm enterprises and they demand a higher level of management skill and technical knowledge (Cunha, 1967). When compared to traditional farm enterprises, the measure of their acceptability is solely economic, this usually being the only criteria on which they are judged and operated (Junge, 1967).

Intensive housing is a part of the practice of intensive animal production. Its development reflects the continuous struggle to reduce the costs of production by reducing investment yet improving working conditions inside the building and aiding in the improvement of biological performance by creating an environment conducive to the optimal conversion of feed into meat. Systems of intensive housing have usually concentrated on the elimination of litter for bedding with a concomitant attempt to increase the efficiency of space utilisation (Baxter, 1974). Housing systems for the lactating sow have not escaped these general trends as has been shown by developments in the UK and other countries.

2. HOUSING PRACTICES IN THE UK IN THE LAST DECADE

There is no really clear picture of the progressive development of housing for the lactating sow and her litter on pig farms in the UK although much can be reconstructed from surveys. In 1968 - 69 for example, information from a survey of 120 pig units in Scotland (Baxter, 1971) suggested the following practices. Whilst the sample then surveyed was not representative of even Scottish units its high proportion of large sow herds provides a starting point against which progress in the subsequent decade can be judged. In the majority of cases sows and litters were housed during lactation in old buildings modified as farrowing units but substandard in ventilation, insulation etc. when compared with the current advisory recommendations (Sainsbury, 1963). A similar situation appears in Norway (Simensen, 1971) and Sweden (Bäckström, 1973) at about this time. The diverse range of housing systems and apparently sub-optimal environments did not however appear to have a significant effect on biological performance. It was concluded that the traditional practice of straw bedding together with the application of sound stockmanship did much to ameliorate otherwise poor environmental conditions. At this time almost all herds practised weaning at 5 weeks of age or older with a third of producers still weaning pigs as late as 8 weeks of age. Immediately before parturition, about 30% of the pregnant sows were housed in individual stalls and the remainder in a mixed collection of pens, yards (covered and exposed) field huts and a scattering of unique combinations. The sow housing of this period was characterised by its heterogeneity. More than half of the pregnant sows however were fed individually, a reflection, of the importance attributed to nutrition and the successful efforts of the nutritional scientists.

During lactation, about one third of the herds were housed for the whole period in one type of accommodation. In the remainder the sows and litters were moved to separate rearing accommodation at some time, usually 2 to 3 weeks, between parturition and weaning, a practice which has subsequently declined with the growing practice of earlier weaning. Eighty seven percent of all farrowing pens contained a crate made from a range of materials but mainly tubular or square section steel. All farrowing pens had solid concrete floors littered with straw or wood shavings and all creep areas were heated, usually with electric heat lamps. Though slatted floors had already been introduced as early as 1957 for fattening pigs (Farm Buildings Association 1957), their application was only just being considered in the UK for farrowing pens (Baxter and Robertson, 1971).

The farrowing pen could still be found in which voluntary farrowing, (solari type pen) Figure 2(v) or only partial confinement during the earlier stages of lactation was practised (swing-side crate and pen) Figure 2(iii). Total confinement in the tubular steel crate (Figure 2(i)) however was on the increase. Farrowing pens averaged approximately 8 m^2 in area for 6 to 8 weeks weaning with crates approximately 2.4 long and 650 to 700 mm wide.

Ten years later the picture has changed dramatically (Meat and Livestock Commission 1980) (Anon, 1980). The trend for the large units of the early 1970s to get bigger has been confirmed, the labour force has declined, investment has increased and the progressive producers have all adopted earlier weaning. The stock of existing buildings has been absorbed and a wide range of new constructions have been built specifically to house the extra breeding stock. Buildings are better insulated and many are now fan ventilated. Individual confinement of pregnant sows in stalls is now practised on 75% of the large herds with about one in four using tethers. Almost all of these

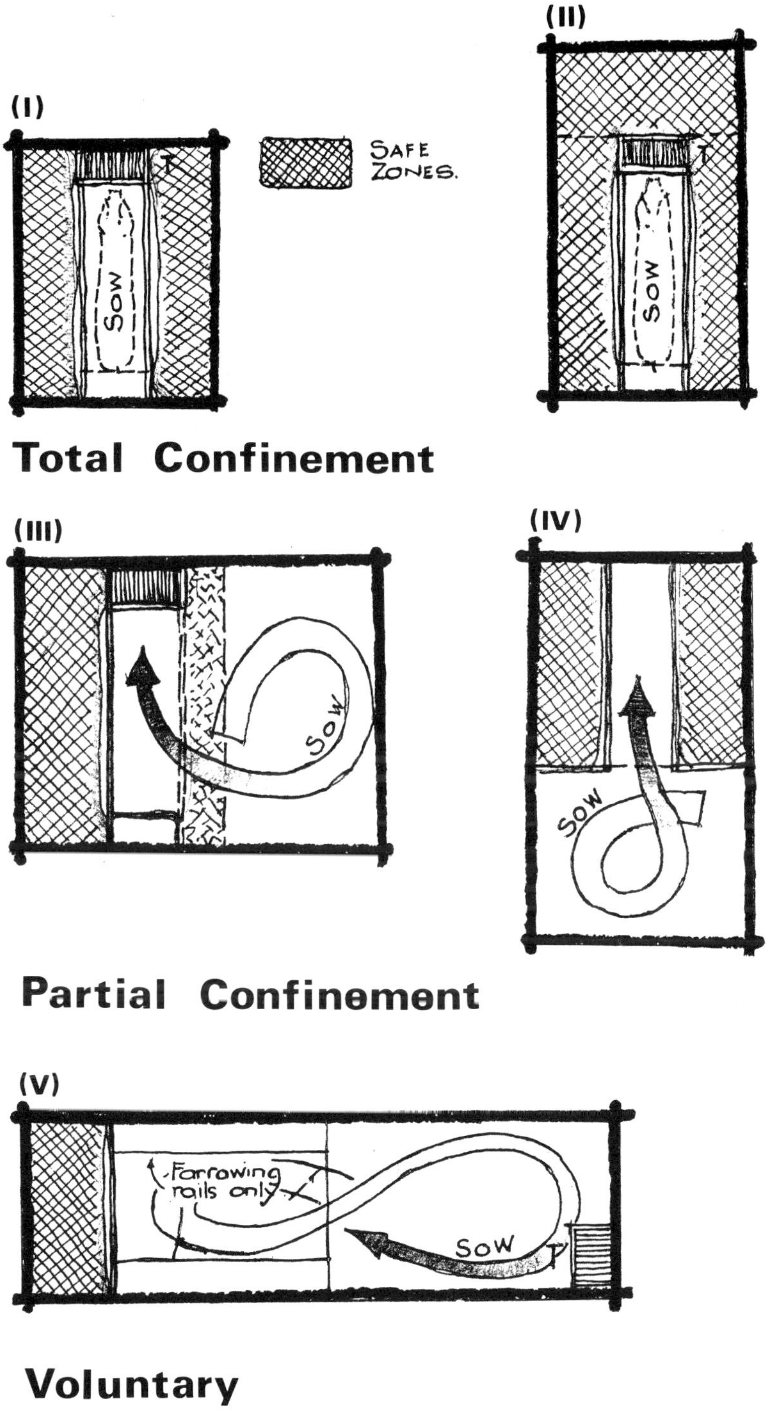

Fig. 2. Farrowing pens.

new gestation buildings contain perforated floors in the rear portion of the sow stalls and few if any use bedding of any type. Individual feeding of pregnant sows is practised on about 85% of all units. In the farrowing house the prefabricated steel farrowing crate is ubiquitous and part or fully perforated floors have become commonplace. Fifty four percent of the latest Scottish sample of 63 herds* appear to have adopted this practice. In farrowing pens perforated floors of mesh or punched metal are more common than those of concrete or steel slats. In sow stalls the reverse is true. The area of the farrowing pen (for 3 week weaning) has now been reduced to 3.5 - 4.0 m^2 and the two layouts shown in Figure 2(i) and (ii) tend to predominate. Further reductions in the dimensions of the farrowing pen by angling the crate appear to be more common in other parts of Europe.

In general the trend is similar throughout the United Kingdom and other parts of Western Europe with some regional variations. In the arable areas of England, for example, sows stalls and liquid manure systems are less prevalent where pens and yards and farmyard manure are still favoured by many producers. Where individual confinement is practised tethers are slightly more popular.

The modern farrowing pen is characterised by its small size (about 3.5 to 4.0 m^2) and the confinement of the sow in a crate approximately 2.4 m long and 700 mm wide. The pen is no longer littered with straw or woodshavings and half or more of the floor surface is in the form of perforated or slatted panels. Though the farrowing room may be kept at a temperature of from 18 to 22°C the baby pigs are supplied with additional heat from electric or gas heaters.

As rapid change in existing systems is clearly not economically, and as a consequence politically acceptable, then the farrowing pen will for some years continue to act as the basic system requiring further understanding and modification.

* Not necessarily the same herds as those surveyed in 1968 - 69.

A detailed analysis of the functional use of space in the farrowing pen is therefore a prerequisite to further development.

3. A SPATIAL ANALYSIS OF FARROWING PENS

One glance at the historical development of pig husbandry shows that of all aspects of pig production, the lactating sow and litter has received the most careful attention. Nevertheless, it is only in the last 30 years or so that the basic design of farrowing pens has changed radically from the ideas of the 18th and 19th centuries or even since Roman times. There are numerous examples in the writings of previous centuries of the description of a farrowing environment similar to that used in the 1950s. The pen for example had to be warm and dry, reasonably spacious with rails next to the walls to prevent crushing of the young pigs. The pen should have a separate area for the litter to feed and rest. Pens of this specification allowed freedom of movement for the sow and litter up to 10 or 12 weeks in an area of about 15 m^2. It contained both an inside and an outside environment.

The design criteria for the spatial zones within a farrowing pen are as follows:

(i) A safe zone for piglets where they can rest and feed if necessary, free from the unwanted attentions of the sow.

(ii) An interaction zone in which the sow and piglets come together primarily for the purpose of suckling although other interactions of importance for the sow's welfare or the piglet's development may also take place here, eg. processes leading to infant recognition, aggressive play between siblings at the udder.

(iii) Space allocation in a farrowing pen must allow the sow to perform (at least) all her maintenance behaviours (feeding, drinking, resting, excreting). This space can be called the sow space. In the case of crate confinement the sow is restricted to one locality within the pen and hence the sow space must be provided within the crate. In part confinement the sow space is not fixed to one locality and the sow may be required to perform different behaviours at different localities. Space must be allocated for this in addition to that required for movement between localities. In the latter case the sow space is dynamic and may overlap with the interaction zone.

3.1. Safe zone for piglets

In Figure 2 some layouts are shown for use in systems of total confinement, partial confinement or voluntary farrowing and the safe zones are indicated.

In layouts (i), (iii) and (iv) the safe zone is only located adjacent to the suckling position of the sow whilst she is confined. In layouts (ii) and (v) additional safe zones have been provided at another location.

The overall size and dimensions of the safe zone will be determined by the spatial requirements of the largest litter. Existing practice indicates safe zones of various dimensions more determined by the overall convenience of pen shape than by the functional needs of the litter. For example in layout (ii) the safe zone at the front of the pen is 0.9 m^2 and has a smallest dimension of 600 mm. This is considered suitable for litters up to 3 weeks of age. In addition, there is the space on either side of the sow amounting in total to another 2.0 m^2 but with a minimum safe width of only 400 mm. In layout (v) for litters up to 6 weeks of age the respective areas and dimensions are 1.3 m^2 and 900 mm; 1.2 m^2 and 350 mm. The latter is not entirely free from intrusion by sow.

In layouts (i) and (iv) the safe zone for pigs up to 3 weeks of age has a total area of approximately 2.4 m^2 with a minimum width of 500 mm. For pigs up to 6 weeks of age the area on one side of the sow would be expanded to give a total safe zone of approximately 4.3 m^2. Layout (iv) provides a safe zone for piglets up to 6 weeks of age of 2.0 m^2 with a minimum width of 840 mm.

The information is summarised in Table 1.

TABLE 1

AREAS OF PENS AND SAFE ZONES FOR A LITTER OF PIGLETS WEANED AT 3 OR 6 WEEKS OF AGE IN PENS OF DIFFERENT DESIGNS

Layout (refer to Figure 2)	Age at weaning (weeks) 3			6		
	Total pen area (m^2)	Total safe area (m^2)	Separate safe zone (m^2)	Total pen area (m^2)	Total safe area (m^2)	Separate safe zone (m^2)
Total confinement						
Layout (i)	4.00	2.40	-	6.00	4.30	-
Layout (ii)	4.50	2.90	0.90	-	-	-
Partial confinement						
Layout (iii)	7.45	-	2.00	7.45		2.00
Layout (iv)	6.15	-	2.40	8.15	4.30	
Voluntary						
Layout (v)	-	-	-	7.65	2.50	1.30

The minimum dimension of the safe areas can vary from 350 mm to 840 mm. Where the dimension is less than the length of a recumbent pig then, whilst the litter will obviously lie in the space available, some lying patterns are restricted. This may in turn conflict with the design of the heated zone of the lying area. For example in Figure 3 the safe zone adjacent to the sow is too narrow for all the piglets to lie safely in the pattern shown yet the extent of the heated zone as determined by the heat lamp does not match this lying pattern. In order to remain safe the piglets will have to adopt a lying pattern whose range extends beyond the limits of the heated area, thereby putting some piglets, usually the smaller, weaker ones at a disadvantage. Alternatively the piglets will adopt a lying pattern, the boundaries of which may extend into the sow space again putting some piglets at risk. A systematic

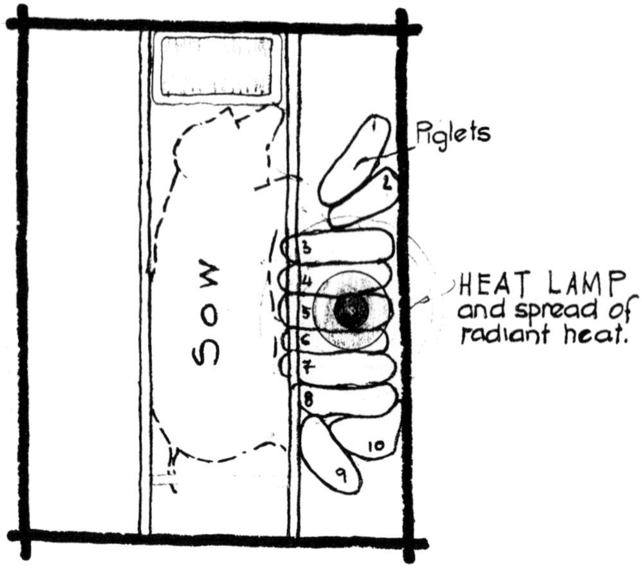

Fig. 3. Conflicting requirements!

appraisal of safe zones for piglets would suggest the following performance specifications.

1. The safe zone must be attractive to suckling piglets of all ages and must be conducive to the piglets' use of it for resting.
2. The safe zone must provide the optimum climatic environment for health and performance as it should be where the piglets spend most of their time.
3. The safe zone must be of sufficient size and of the correct shape to accommodate all piglets in their preferred huddling/grouping arrangements at all times.

Considering the implementation of these performance specifications highlights a number of problems and suggests many topics for further study.

To begin with, it has in the past proved inordinately difficult to find ways of attracting piglets away from the udder in their first 48 hours of life. It has been suggested that there are a number of very strong motivational processes which serve to keep the piglets at the udder and up until now no agricultural practices have been able to overcome these piglet motivations. However by investigating the nature and control of these motivational processes in addition to studying the general cognitive development of the piglet we could deduce what is required to attract the young piglet away from the sow. Whether it would prove commercially feasible to provide such an attractant would depend on its nature.

The probable reason for so little attention having been given to the design of safe zones for piglets is the lack of information on the piglets' requirements.

To be able to design a safe zone for piglets which will be conducive to their welfare as well as their productivity we need to be able to answer the following questions.

1. How does the size and shape of space required by piglets vary with their size, the size of litter and the climatic environment?
2. What are the required characteristics of the safe zone environment?

3.2. Interaction zone

Clearly in any space containing both sow and piglets the smaller animals may be in some danger from the actions, accidental or intentional of an animal outweighing them by a factor of 100 times.

It will however be impossible to prevent all dangerous situations from developing as many of these are unpredictable and in some cases unrepeatable. However, certain activities and some places in the farrowing pen are known to be associated with potential risk to the piglets. In an interacting situation between sow and piglet(s) the high risk sow activities are:

(i) The sow lying down

(ii) The sow standing up

(iii) The sow moving about in the pen.

In the transitional movements between lying and standing and vice versa, the danger of crushing piglets is compounded by the proximity of walls or other unyielding hardware against which a piglet may be trapped. This is particularly unfortunate as piglets generally prefer the apparent convenience, comfort and/or security of lying against or close to walls etc. Other impediments to the rapid escape movements of the piglet such as smooth or irregular floors may also increase the potential risk. All dangerous situations may be worsened if the piglets are not easily seen or sensed by the sow. In all movements by the sow likely to lead to piglet injury, the danger is increased if the sow acts suddenly and violently. Slow, gentle, deliberate movements of a healthy sow are readily tolerated provided that enough suitable space for the movement is allowed.

It may be thought that standing up by a sow is a simple reversal of the lying down process and consequently would give rise to similar dangers. There are however fundamental differences between the movements involved in standing and lying and these give rise to completely different dangers. Lying down proceeds as follows. After selecting the site for lying (if unrestrained) the sow will adjust her stance and her body posture in ways that appear to maximise her stability. She lowers herself onto her front knees and then slides one knee under her body so that she is resting on one knee and one shoulder. In a series of squirming movements she increases the rotation of the front half of her body until it is in an almost recumbent position. She then gently drops her hindquarters to the ground and adjusts her posture to either of the typical sternum or recumbent lying positions.

It appears as if this basic pattern of movements is adaptable to different environmental conditions. For example in certain crates the rotational movements of the front part of the body are reduced with the consequence that the hind quarters flop to the ground more violently than normal. Sows observed in large straw bedded pens were seen to alter their lying down behaviour when very young piglets were around. Instead of dropping their hindquarters onto their rear flank they bent both hind legs together in the way humans sit down (Baxter, 1980a).

What happens to the sow's hindquarters seems to be the most important aspect of lying down. The rapidity and abruptness of these movements could and indeed have been seen to result in piglets being trapped against bars or rails or even crushed against the floor.

In standing up, the dangerous movements appear to take place at the beginning and final stages of the sequence. From a relaxed recumbent position the sow must rapidly shift her centre of gravity over her legs to the sternum position so that she can then raise her body. This is normally done by lifting and twisting the head and shoulders whilst kicking with the legs. Once the fore-end of the body is upright the rear-end is twisted into position. These violent kicking and twisting movements may result in piglets being propelled against rails or walls with sufficient force to cause injury or even death. If there are any obstruction e.g. rails or walls, in the path of these kicking movements there is a high risk of piglets being trapped and crushed during standing up by the sow. Space restrictions which tend to trap the sow under rails etc. increase the frequency and magnitude of violent movements as the sow attempts to release herself. Once in the sternum position the sow alters her posture in preparation for raising herself into a 'dog sitting' stance. This she achieves by 'walking' backwards with her forelegs whilst her hindquarters remain on the ground. After a pause of highly variable duration the sow will rise to a quadripedal stance using a jerky, forward-lunging of her head to counterbalance the raising of her hindquarters. The final components of standing up behaviour involve taking one or two steps forward with her forelegs to adjust her balance. This final component of standing may involve a risk of trampling on piglets. As a balancing movement these steps forward will not be easily modified or stopped to avoid an over-inquisitive or lethargic piglet.

General locomotor behaviour would appear to cause few problems unless space is severely restricted or the sow is free to move around in a voluntary farrowing pen. In such cases violent or hurried movements resulting from muscle fatigue or injury caused by confinement or simply due to the sow galloping around in the free situation, may result in trampling underfoot or crushing against pen components. A suggested performance specification for the interaction zone would be as follows:

1) The sow's environment should encourage her to stand up and lie down with a minimum of violent and sudden movements and in a way which minimises the risk of traumatic injuries to the piglets.

2) The duration of the interaction between sow and piglets should be minimised.

In order to implement these specifications through better design, answers to the following questions are needed:

1) What are the nature of the interactions between sow and piglets?
   (i) In particular what is the nature of the piglets' motivations to interact with the sow? (this raises the possibility of allowing the litter to satisfy these motivations in a safe zone).

2) What are the movements that the sow makes in standing/lying which are likely to injure the piglets?
   (i) How could the space in which these movements occur be modified to eliminate or ameliorate their injurious effects?

3.3. The farrowing crate (sow space)

The previous discussion on the standing and lying behaviour of the sow was directed to a consideration of the piglets' interests. It is important also to consider the sow's requirements. Resting demands that the sow adopts a lying posture whilst the consumption of food and water, the voiding of excreta, and the visual exploration of her surroundings demand an upright stance. Consequently the transitional movements of standing up and lying down are essential to the sow irrespective of the environments in which she is housed. It is important to the sow, not merely for reasons of her psychological wellbeing, that she should be allowed to perform these behaviours in her preferred manner and without discomfort. In contrast to the many bovids which have to rise hindquarters first in standing up or the elephants which have evolved knee structures, the relative mass of the mature sow has led to no particular evolutionary specialisms in her methods of standing and lying. Nevertheless, the movements incurred in standing up or lying down are likely to have precise mechanical significance in the vertical movement of up to 200 kg over a distance of 750 mm. This suggests that small impediments to any of these transitional movements may impose severe mechanical as well as psychological stress upon the sow.

In addition, the sow needs space for feeding and drinking. Although we may now only be concerned with the movements and spatial requirements of part of the sow's anatomy e.g. head and neck, injuries to head and face, and decomposing feed in troughs provide some evidence of neglect in this area.

Although considerations of the biological requirements for sow space are paramount, it would be naive to assume that there were no attractions for management in the confinement of sows. As a management aid, the crate contains the sow and limits her actions and position within the building. In this way the piglets can be contained in an enclosure which is simpler to construct and is consequently less expensive. The sow is easier to control, safer to handle, and less likely to interfere when the piglets are being handled e.g. teeth clipping, ear tagging, injecting or castrating. Though, as a result of close confinement the sow must defaecate, urinate and lie in the same place, the development of perforated floors has facilitated the removal of much of the excrement and has helped in the provision of a relatively clean and dry lying area.

Though the idea of the crate may have been originally conceived as an apparatus to protect piglets from crushing, it is possible that its current value lies more with other managerial advantages. However, until successful alternatives to the crate are available, it will continue to be the focus of attention in farrowing pen design.

The performance specifications for the sow space are as follows:

1) The sow space must be large enough and of the correct geometry to allow the sow to stand up and lie down in her preferred manner and with a minimum of discomfort.

2) The feeding and drinking equipment should allow the sow comfortable and unimpeded access to food and water. Furthermore, the shape of the food trough should allow the sow to reach all the food in it.

To be able to design a sow space we need to be able to answer the following questions:

1) Which movements of the sow's preferred standing and lying behaviours are required mechanically and what space do they require?

2) How much bigger, if at all, is the sow's psychological representation of her need for space?

3) What is the size and shape of a sow's head?

4) What space is required for the movements which accompany the ingestion of food and water?

4. AN ANALYSIS OF THE EFFECT OF FLOORS ON ANIMALS IN FARROWING PENS

The floor or ground is an important feature of the pigs' environment. When compared with the carpet of leaves in a forest or the grass in a meadow the concrete floor of the modern piggery is unimaginative and unyielding. To the pig, natural floors provide more than just mechanical support, they are a readily manipulated source of food and novelty. As a mechanical support system they are stable yet resilient, organic and interesting, but they are also the ideal environment for organisms which may be harmful to the pig. Natural floors are multi-functional and adaptable.

On the other hand the present artificial concrete floor of the piggery is hard and unyielding, resisting rather than absorbing the dynamic interplay of forces between floor and animal, demanding that the animal itself should cushion all the effects of its physical environment. Under many circumstances, the body of the pig copes, but the level of tolerance to extremes of impact or abrasion are lowered by the hostile nature of the floor materials. Yet the concrete floor satisfies many of the requirements of management. Its flat, smooth, durable nature is ideally suited to the activities of man and his wheeled appliances. Though not totally inhospitable to the needs of micro-organisms the concrete floor can be readily and easily cleaned and disinfected. Its durability of shape and structure outperforms the natural floor but therein lies the dilemma. The very characteristics of the floor suited to the pigs' needs are antagonistic to those of management. Clearly there is a long way to go in the design of floors which satisfy both man and animal.

There are a number of factors which interact and influence decisions on the choice, design and operational characteristics of floors (Figure 4).

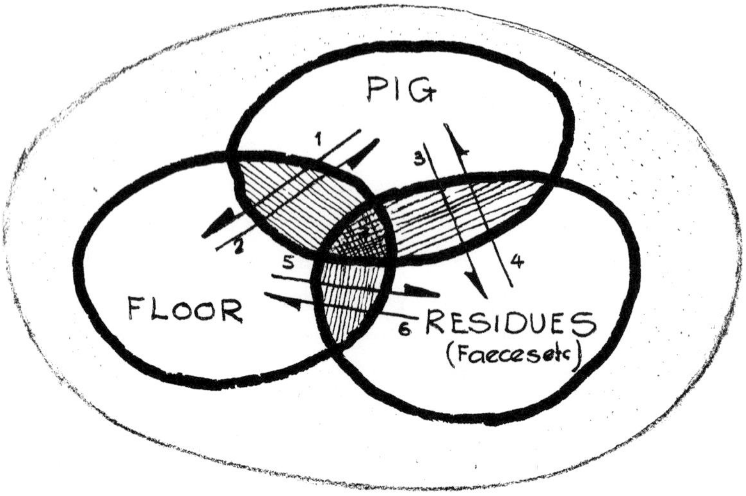

Fig. 4. Floor interactions

Changes have to be instigated when the result of these interactions is incompatible with the requirements of management. For example, when the effect of the floor on the pig is injury or the effect of the residuals on the pig is dirty pigs then management must intervene and seek alternatives which will reduce or eliminate these inconsistencies.

If we consider the traditional concrete floor covered with straw bedding, we find that it is mainly the bedding which reduces most of the inconsistencies. It reduces the conductive heat loss from the pig, it acts as a partial cushion between animal tissue and floor and it contains the faeces and some urine, reducing some of the various floor/residue interactions whilst still maintaining relatively clean, dry pigs. However, when both litter and labour become expensive the systems based on bedding and farmyard manure become financially incompatible with the economic objectives of management. A solid concrete floor without litter may be part of the answer especially if the design of the pen layout takes advantage of the pigs discriminate excretory behaviour. The fact that the pig uses space selectively for resting and elimination allows the model in Figure 4 to be considered in two parts with two floor solutions. This is clearly demonstrated in layout (iv) Figure 2 where space is differentiated into resting/feeding area and dunging space and where the latter floor can be unbedded and/or uninsulated. Even in this system however, the pigs excrement has to be removed from the immdediate vicinity of the pig - an obnoxious and time consuming task. It is little wonder that the perforated floor was developed.

4.1. *Perforated floors

The perforated floor is essentially a labour-saving device and despite the fact that it costs more to install than a solid floor it can reduce or eliminate the task of muck handling in the piggery. Soaring wages prohibit the use of labour on all but essentially productive jobs and so, with decreasing availability, good quality labour demands good working conditions. The economically hard pressed producer has little choice in such a situation, especially when it has been demonstrated in some studies, that pigs reared on well designed, installed and maintained perforated floors can, without injury, grow as fast on as little food as those on other systems. Experience too, has shown that concern for animal welfare is valid for there are many instances of animals injured and deformed as a result of being housed on bad perforated floors.

The main characteristic of the perforated floor is the incorporation of voids through which excrement can pass. The selection of dimensions for solid areas and voids in a perforated floor is of major importance. To be safe, no perforated floor should allow an animal housed on it to put its foot through the perforation, for although the foot can sometimes be withdrawn without injury to the pig, broken legs, cuts, lacerations and abrasions are a more likely occurrence. This is more important with heavy mature stock than with young, lightweight and mobile pigs. With large stock not only should the

* A generic term to cover all forms of slatted, sparred, mesh or pre-punched floors.

foot be unable to pass through the openings, but to make sure that the animal can place its foot anywhere on the floor with relative ease and confidence, it would seem reasonable to assume that the width of the perforation should not exceed half the width of the narrowest dimension of the foot's normal contact area with the floor.

There is some evidence available which provides us with a relationship between minimum foot dimension and bodyweight in LW x L pigs (Mitchell and Smith, 1978). More extensive reliable data is needed to substantiate the following expression but its use in this discussion may prove useful.

Minimum width of foot (mm) = $10\ w^{0.33}$, where w = bodyweight (kg)

Figure 5 shows the predicted maximum width of perforation based on half the minimum foot width for pigs of liveweights up to 200 kg. The data seem to be supported by practical experience where sows and gilts in stalls are usually on 25 mm slots, grower/finisher (45 to 90 kg) on 18 mm slots and baby pigs on 10 mm gaps.

Where however, due to lodging of faeces in very narrow openings, it is not practical to use the smaller sized perforations, an increase in slot width should be accompanied by an increase in the width of the solid area in slatted floors. Increasing the ratio of solid to void ensures less chance of the animal standing over a void.

Apart from the physical characteristics of the faeces and the amount of animal movement, lodging of faecal material in the perforations would appear to be a function of the complete geometry of the apertures and the rigidity of the floor membrane.

Where for example, 75 mm deep concrete slats are placed only 10 mm apart then lodging is almost certain to occur, while similar apertures in metal sheet with a depth of only

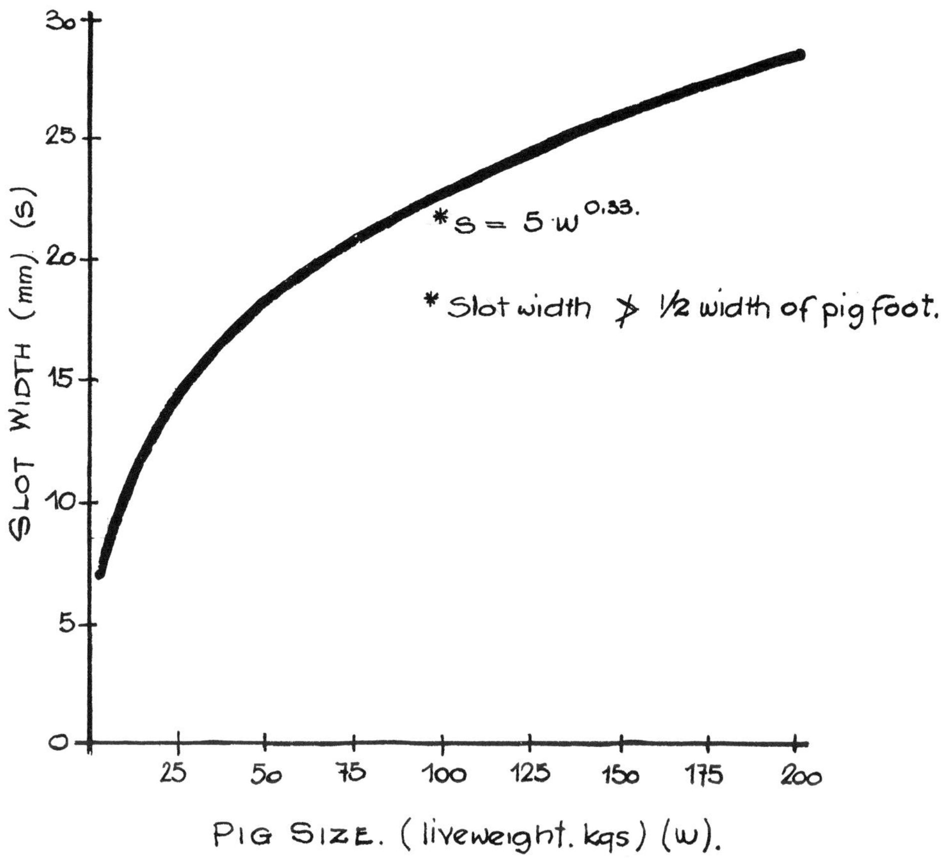

Fig. 5. Relating slot size to pig size.

2 mm will remain clean. Narrow concrete slats approximately 50 mm wide are usually more prone to lodging than slats 100 mm wide, for in order to ensure a comfortable floor narrower slots would be used and the sides of the concrete slat must be perpendicular to the face in order to provide adequate cover to the steel reinforcement. Narrow slots bounded by deep parallel faces of rough texture are prone to bridging and lodging with faecal material. Where solid slats are used it would appear that they should have a sideslope of at least 1 in 5 and a surface width of not less than 4 times the width of the slot.

Mesh floors, another category of perforated floor, are usually constructed of welded or woven rods or expanded metal and require different considerations. Mesh floors are characterised by their high percentage of voids, by the narrow width of the solid areas and the shallow depth of the floor membrane.

Compared to 100 mm wide concrete slats and 25 mm slots where the percentage void in the floor is approximately 20%, the voidage in mesh floors is nearer 60%, so it is little wonder that the floor keeps clean even when the width dimension of the hole is the same. However, due to the thin and rather sharp nature of the floor membrane, it is even more important that feet and toes do not get caught in the perforations. Although a flexible membrane may also help to dislodge the faecal material in the perforations, the whole floor must be adequately supported to eliminate excessive flexure resulting in metal fatigue and uncomfortable stock. All edges of narrow membranes must be clear of burrs and sharp spots.

Where there is great variation in the foot size of animals kept in the same pen e.g. in farrowing and sow and litter pens, some form of mesh floor is likely to provide the best solution. For example, the dimensions of a mature sows' foot may exceed that of its piglets by a factor of 4 or 5 and this cannot be readily accommodated in a single slot width, slatted floor.

The performance specification for a good floor, whether solid or perforated should include the following (Baxter and Mitchell, 1977):

1) It should not contribute to injury.

2) It should not contribute to disease.

3) It should not contribute to ill-health or discomfort.

4) It should not contribute to inconvenience.

Some considerations of the above factors follow.

4.2. Injury

There is adequate evidence to show that the interaction between floors and animal tissue often results in traumatic injuries of varying degrees of severity (Penny, 1978; Smith, 1978). This problem is being studied in a number of countries. Much of the work however, is concerned with the evaluation of specific floors and materials (Smith and Mitchell, 1976) and much less effort seems to be directed to more basic questions. At what point, for example, on the stress/strain curve does injury occur for each tissue? Though Smith and Mitchell (1976) have described injuries resulting from perforated floors such as bruising, erosion and haemorrhaging no analysis has yet been presented on the relationship of these to the separate characteristics of floors.

It is necessary to re-emphasise that the results of animal/floor interactions do not rest solely on the characteristics of the floor - even the worst floors will cause fewer traumatic injuries if the pig does not move over the floor. Most traumatic injuries result from the dynamic interaction of pig and floor. Cuts and abrasions would appear to result from similar sliding motions but are differentiated by characteristics of the floor. Bruising and puncturing may be caused by similar forces but entirely different floor properties.

With static forces, the puncturing of the skin arises from a high intensity of pressure over a very small area - a sharp point for example. With a constant load, as the area increases the resulting injury may change from puncturing to bruising. The structural integrity of the body tissue may be sufficiently strong for it to carry the load between support points without rupturing especially if the supporting points are organised to carry the load in a well distributed fashion.

On the other hand, cutting usually results from the sliding motion of body tissue across a point or line of small area but high pressure. The depth and degree of isolation of the sharp point or edge will influence the type of cut inflicted. Abrasion seems to be the result of the body tissue being moved across a multitude of small cutting points or edges. For the same dynamic force the abrasing surface creates more resistance than an isolated point or edge and usually results in wounds of smaller dimensional characteristics. From a dynamic point of view, less movement of either the whole body or individual limbs by the pig will result in less cutting and abrasion. This is not likely to play a significant role in reduction of injuries due to static forces. Injuries due to both static and dynamic conditions, can only be reduced by eliminating or reducing points or edges of small area and high pressure intensities.

Unfortunately, if we tried to eliminate all resistance to dynamic forces we would end up with a floor on which the animal could not walk but would merely slip and slide, so the surface of the floor must provide some resistance to motion. All floors at the moment do this by providing a surface of small points or edges but other possibilities might include properties concerned with adhesion or suction for example.

This aspect of animal welfare is sufficiently obvious and serious as to warrant immediate attention. Answers to the following questions are urgently required:

1) What components present in floors cause what sort of injuries?

2) What radius of curvature must there be on an edge so that cuts are minimised?

3) How soft should a floor be to prevent bruising or impact injuries?

4) How firm or rigid should a floor be in order to support an animal securely?

5) At what point on the stress/strain curve does injury occur for each tissue?

6) What rate of abrasion is produced by different floors and what is tolerable?

7) What dynamic stress is necessary for foot tissue and bone in order for it to maintain its mechanical strength?

8) Is bruising physiologically necessary in the feet of piglets before hardening of the tissues can occur?

9) What fraction of the area of an animal's foot can support its whole weight without injury?

10) Can we model the stress/strain interaction between feet and floor, using engineering methods, and hence predict the effect of novel floor types?

4.3. Disease

Floors contribute to disease, by providing an environment for the survival of pathogenic organisms and a source for access to the animal. They may also create the portal of entry for the organism by creating injuries which penetrate the boundary tissues allowing ingress to the blood stream and subcutaneous layers. Floors create a suitable environment for bacteria when they create a suitable substrate for survival - when they provide a source of food, oxygen (in some cases) and thermal conditions conducive to growth and reproduction. No floor can ever be kept free of some organic substrate but the intention must be to limit the conditions for the growth of micro-organisms and so reduce the likely challenge to the animal. Cleaning reduces the available substrate only on those parts of the floor exposed to the cleaning medium which in many cases will not penetrate to sub-surface levels.

Disinfectants, where they penetrate to operative levels, either kill the micro-organisms or limit their growth. Sterilisation usually by inducing high temperatures e.g. flame gun, attempts to kill all organisms but only does so if the temperature is high enough at organism level. Another method would be to provide an impermeable layer at intervals between the living micro-organisms and the pig either around the pig or around the organisms. Whatever the technique used the aim is to reduce the potential challenge of pathogenic organisms by elimination, population reduction or separating the two environments at regular intervals of time.

Disease is part of a constantly changing dynamic system and is not easily reduced to a number of specifiable requirements. Nevertheless, some answers are needed to questions of the following type:

1) What is the rate of growth of micro-organisms in the general floor environment of both solid and perforated floors?

2) What are the dynamics of micro-organism populations with respect to disinfecting procedures and the systematic depopulation of pens?

3) Is there any correlation between the amount of faecal and other decaying organic matter in a pen, the population of pathogenic organisms, and the incidence of disease?

4.4. Distress and discomfort

Whilst a resilient floor may provide comfortable conditions for resting and safe conditions for walking, excessive deflections occurring suddenly and unpredictably are likely to result in some distress. This is more likely to occur on suspended perforated floors than on solid floors. Where the floor consists of a number of separate members like slats, excessive deflection of one member adjacent to a non-deflecting member results in an uneven floor. Uneven floors, however they are caused, may be unsafe for locomotory activities and uncomfortable for resting behaviour. The latter is especially true where the point of contact with the animal's anatomy is in an area with little covering of cushioning muscle. Floors which are noisy when the animals move on them, such as light metal slats, may under some conditions give rise to conditions of surprise and uneasiness.

Many floors also result in thermal discomfort, especially to the resting pig, where the surface temperatures of contact areas are reduced and the body loses heat to the surrounding floor. Solid floors with no insulation or where the insulation layer is embedded under a thick surface layer of concrete create the worst conditions and floors covered with straw bedding, the most satisfactory conditions. Perforated mesh floors generally have a neutral value, having no more effect on the resting pig than the surrounding air has on the standing animal (Bruce, 1977). Wet floors or floors covered with slurry will increase heat losses in cold environments, but may be beneficial in hot climates. The $R_{f_{45}}$ as suggested by Bruce (1978) is shown in Table 2 for a range of floor types.

TABLE 2

RANKED CLASSIFICATION OF FLOORS BY THERMAL RESISTANCE

Floor type	Thermal resistance $R_{f_{45}}$ ($^{o}Cm^2/W$)
60 mm dry straw on concrete	0.66
25 mm dry straw on asphalt	0.52
25 mm dry sawdust on concrete	0.23
Wooden slats 58 mm wide, 70 mm deep, 10 mm gap	0.23
18 mm cement screed on 150 mm lightweight aggregate	0.17
Metal mesh floors	0.12
Slatted and 'T' bar pvc slats	0.11
Concrete slat panel	0.086
Very rough concrete floor with slurry	0.057
Muddy ground	0.044
Uninsulated concrete floor	0.042

Much is now known about the thermodynamics of heat loss between an animal and the floor on which it rests, sufficient indeed for the thermal characteristic of the floor to be well defined by the $R_{f_{45}}$ value. On the other hand, the most efficient method of providing the necessary thermal environment for piglets, an essential ingredient in the definition of the safe zone, is not well understood. Questions related to the use of heated floors, or air heaters cannot yet be effectively resolved until the following are answered:

1) What is the piglets' threshold of tolerance with respect to floor/air temperature difference during resting?

2) What are the piglets' responses to alternative methods of providing thermal conditions in
   (i) the safe zone,
   (ii) the interaction zone.

4.5. Inconvenience

The inconvenience of a floor is measured in management and financial terms. Solid floors which take a long time to clean and disinfect or which do not clean satisfactorily, absorb labour. Floors which cause damage to equipment or lead to rapid deterioration of scraper blades are costly to maintain. Floors which deteriorate rapidly due to corrosion or wear are not only costly to maintain but can lead to animal injuries. The perforated floor was developed to reduce the inconvenience of having to clean regularly organic residues from the floor surface.

In this respect perforated floors which retain faecal material both on the floor and in the interstices are not functioning adequately and additional cleaning leads to inconvenience.

5. BEYOND CONFINEMENT?

Although rapid changes in farrowing pen designs are impractical due to economic constraints, there is no reason to believe that fundamental changes should not be long term objectives in agricultural research. The need for such change is already suggested from three different angles. Firstly, lay members of society express great concern over the suffering the sow appears to go through whilst confined in a farrowing crate. The representatives of these members of society, the campaigning welfarists, are beginning to have considerable political weight in many European countries. The public dismay at such intensive livestock practices as confining sows at farrowing supported by the threat of prohibitive legislation may generate an urgent need for major changes in farrowing pen design.

Secondly, a number of recent studies have brought into question the effect of restraint at farrowing on sow productivity. Although none of the work has been carried beyond a preliminary stage, workers at SFBIU (Baxter and Petherick, 1980) have suggested a way in which restraint prior to farrowing may increase the incidence of still-births. Workers in Sweden have highlighted a possible effect of restraint on the incidence of Mastitis-Metritis-Agalactia Syndrome.

Thirdly, the importance of early environment influences upon the normal development of an animal's biology is now beginning to be realised by agricultural scientists. Although such influences have long been acknowledged in developmental biology and psychology, it was only the recent discovery of their importance in the development of boar sexual and social behaviour that has brought them to the attention of the agricultural community. With such information available in an agricultural context, it should not be long before the effect of the environment upon the development of piglet biology and behaviour comes under wider scrutiny. It is likely that this scrutiny would relate to both their productivity and their welfare.

Considerations such as these raise doubts about such fundamental issues in crate farrowing that the entire rationale upon which conventional farrowing pens are designed and used needs to be questions. Crate farrowing is also an 'unnatural' method of animal exploitation inasmuch as it attempts to suppress rather than exploit the animal's own biological adaptations. For example, although the main objectives in the farrowing pen design are the provision of a safe (from crushing) and climatically suitable environment for the piglets, restraining the sow in a crate prevents her nest-building, the functions of which would appear to be mechanical and climatic protection (Gundlach 1968).

It is perhaps time to reconsider the design of farrowing accommodation by a fundamental study of the environmental conditions required by the sow and litter for optimum performance and welfare. Conjectures about the nature and feasibility of such fundamental approaches to the design of livestock environments have been documented over a number of years (Baxter, 1975; Rogerson, 1971). Developed from concepts of occupant-oriented design conceived for human architecture it has been suggested that livestock environments should be designed on the basis of systematically deduced environmental requirements of livestock. In the terms of the systems model outlined in the introduction, livestock environments should be designed on the basis of the requirements of the biological system rather than the managerial system. This concept of occupant-oriented design has been discussed in some detail in relation to the study of animal behaviour in a recent paper by Baxter (1980b).

On this basis, a re-evaluation of the environmental requirements of sows and litters reveals the following.

The potentially harmful effects of restraint previously mentioned would appear to be caused by the frustration of the sow's nest building behaviour. The accommodation of nesting motivation would therefore seem important both for sow productivity and in response to public concern over sow welfare.

The easiest way to accommodate sow nesting motivation would be to allow the sow to perform the requisite nesting behaviours. From preliminary observations of sows in large straw bedded pens, nesting behaviour appears to proceed in two stages. Firstly, the sow moves around in an apparent state of high arousal gathering straw and piling it in one corner of the pen. Secondly, the sow will root, paw and burrow into the resulting large pile of straw eventually forming it into a cup shape. This second stage will end a short while before the birth of the first piglet. From this the sow's two environmental requirements for the performance of nesting would appear to be space and the provision of a nesting material.

This is not the only possible method of accommodating nesting motivation. It may prove feasible to provide the sow with the environmental stimuli which satisfy nesting motivation prior to or at the same time as she begins nesting behaviour. If sows prove to have similar motivational control of nesting as birds do then on the basis of Hinde's (1958) study of canaries we would predict that the provision of a completed nest in the sow's pen would considerably reduce her nesting motivation.

Although highly speculative, this discussion should make clear the importance of an understanding of the motivational control of nesting in the design of farrowing accommodation.

In the same studies of sows the litters unrestrained in large straw bedded pens (at SFBIU) a number of other sow and piglet behaviours were identified which appeared to have the specific function of preventing injury to the piglets. For example, the most remarkable of these was the sow's lying down behaviour. The sow had delivered her piglets in large and luxurious nests to which both sow and litter returned to rest for a number of days *post partum*. When approaching the nest to lie down, the sow would drop onto her forelegs just outside the nest and move forwards rooting the straw to the front and to the side of her. Once completely within the nest perimeter she would drop her hindquarters gently to the ground. On many occasions the sow's rooting movements uncovered a sleeping piglet and removed it from the danger area. Consequently the function of this behaviour is concluded to be piglet protection.

With pre-weaning mortality of piglets constituting one of the major losses in pig production, any occupant-oriented design of farrowing accommodation must proceed with a thorough understanding of all of the sow's and the piglets' piglet protection behaviours.

Speculation on the issues raised by these matters suggest that questions like the following need to be answered:

1) In general, what is the nature of the interactions between sow and litter and what types of movements are typically involved in these interactions?

2) Specifically are there any behaviours in either the sow's or piglets' repertoire which have the specific function of piglet protection? If so what are the environmental conditions required for their adaptive performance?

3) What is the nature of the motivational control of nesting behaviour?

6. CONCLUSIONS

This paper has directed its attention to asking questions rather than providing answers. As such its direction is towards the future. Although an attempt has been made to describe the present situation in sow and litter housing in the UK and to reflect briefly on the previous trends of the decade from 1970 until now, this has been done to provide a springboard for future considerations.

This paper has also concentrated on those moves which may have to be made or the questions which will need answering in response to a growing concern for animal welfare. The paper has concentrated on two aspects of farrowing house design - space and floors reflecting the major welfare issues of confinement, injury, distress and discomfort. The questions that have been raised, fall into two classes - those with an immediate appeal and those which fulfill a longer term and more basic requirement for information. In the former case, the present farrowing crate and pen is analysed in order to seek improvements whilst in the latter situation alternatives to the crate may need to be found.

Whatever questions are eventually asked and whatever answers are gathered the whole picture will remain unclear until they can be adequately located within an acceptable, all-embracing, well-defined concept of welfare.

REFERENCES

Anon, 1980. Who Does What in Pig Production? Brit. Frmr. and Stockbreeder, 5 July 1980, 36-37.

Bäckström, L., 1973. Environment and Animal Health in Piglet Production. A Field Study of Incidences and Correlations. Acta. Vet. Scand. Suppl. 41, 1-240.

Baxter, M.R., 1980a. Personal communication.

Baxter, M.R., 1980b. Environmental Design for Animal Production: The Role of Ethology (in press).

Baxter, M.R. and Petherick, J.C., 1980. The Effect of Restraint on Parturition in the Sow. Proc. Inter. Pig Vet. Soc., 1980 Congress. June/July Copenhagen. Nielsen, N.C., Høgh, P and Bille, N. (Eds), 84.

Baxter, S.H., 1971. A Survey of Management and Housing of Lactating Sows and their Litters in Scotland. Frm. Bldg. R. & D. Studies (01) May, 1971 3-15.

Baxter, S.H., 1974. Intensive Housing and Its Effects on Farm Animals. S. Afr. J. Anim. Sci. 4, 259-264.

Baxter, S.H., 1975. Farm Buildings Design and Layout. Paper to Joint Consultative Organisation for Res. and Dev. in Agric. and Food, Engng. and Bldgs. Board. Frm. Bldgs. Commt. (unpublished, April 1975).

Baxter, S.H. and Robertson, A.M., 1971. Slatted Floors in Farrowing and Rearing Pens. Frm. Bldg. Prog. (26) October 1971, 27-36.

Baxter, S.H. and Mitchell, C.D., 1977. Developments in Floor Construction in Animal Production. Vet. Annual 17th issue 1977, 286-291.

Blaxter, K.L., 1967. Nutritional Problems Arising in Intensive Animal Production - Energy Aspects. 9th Int. Congr. Anim. Prod. Edinburgh 1966, 73-79 Edinburgh Oliver and Boyd.

Brambell, F.W.R., 1965. Report of the Technical Committee to Enquire into the Welfare of Animals Kept under Intensive Livestock Husbandry Systems. Cmnd. 2836 December 1965 London HMSO.

Bruce, J.M., 1977. Conductive Heat Loss from the Recumbent Animal. Frm. Bldg. R. & D. Studies (08) February 9-15.

Bruce, J.M., 1978. A Suggestion for a Ranked Classification for Floors of Livestock Buildings Based on the Conductive Heat Loss from Recumbent Animals. Proc. Symp. 'Animal Housing-Injuries Due to Floor Surfaces'. November 1978, 123-131. Cement and Concrete Association, Slough, England.

Cunha, T.J., 1967. A General Survey of the Problems of Intensive Animal Production. 9th Int. Congr. Anim. Prod. Edinburgh 1966, 41-45, Edinburgh Oliver and Boyd.

Farm Buildings Assoc., 1957. Farm Visits Report: East Nevay, Eassie. Frm. Bldgs. Ass. J., December 1957, 10.

Gundlach, H., 1968. Brutfürsarge Brutpflege, Verhaltensontogenese und Tagesperiodik beim Europaischen Wildschwein (Sus Scrofa L). Z. Tierpsychol., 25, 955-995.

Hinde, R.A., 1958. The Nest Building Behaviour of Domesticated Canaries. Proc. Zool. Soc. Lond. 131, 1-48.

Junje, H.R., 1967. Intensive Animal Production-Problems of Environment. 9th Int. Congr. Anim. Prod. Edinburgh 1966, 96-103, Edinburgh Oliver and Boyd.

Meat and Livestock Commission, 1980. Pig Improvement Services: Regional Hit Count. March 1979. Data Sheet 80/3 Bletchley, England.

Mitchell, C.D. and Smith, W.J., 1978. Piglet Foot Dimensions for Design of Slotted Floors. Frm. Bldg. Prog. (51) January 1978, 7-9.

Penny, R.H.C., 1978. Genetical, Phsyiological and Anatomical Factors Contributing to Foot, Limb and Other Disorders, Attributable to Floors in Growing and Adult Pigs Including A Statistical Review of Foot and Limb Disorders in Pigs Attributable to Floors. Proc. Symp. 'Animal Housing-Injuries due to Floor Surfaces'. November 1978. 29-47, Cement and Concrete Assoc. Slough, England.

Rogerson, P.D., 1972. The Size of Cattle and Their Requirement for Space. Frm. Bldg. R. & D. Stud. (03) November 1972, 3-18.

Sainsbury, D., 1963. Pig Housing. Ipswich Farming Press (Books) Ltd.

Simensen, E., 1971. Miljøfaktorer i Svine produksjonen Meld. No. 60 Inst. Bygningsteknikk, Norges Landbrukshøgskole.

Smith, W.J. and Mitchell, C.D., 1976. Observations on Injuries to Suckled Pigs Confined on Perforated Floors with Special Reference to Expanded Metal (2073F). Pig Vet. Soc. Proc. 1 October 1976, 91-104.

Smith, W.J., 1978. The Effects of Environmental Factors, Floor Design and Materials on Foot and Limb Disorders in Baby Pigs. Proc. Symp. 'Animal Housing-Injuries due to Floor Surfaces'. November 1978, 49-56. Cement and Concrete Assoc. Slough, England.

DISCUSSION

A. Hoogerbrugge *(Netherlands)*

You postulated that if sows were crated during pregnancy then the time of parturition, the total duration, was increased. We have made observations on more than a thousand parturitions in Holland and never found this. It might be a genetical predisposition that one breed is more susceptible than others, but on average we had a parturition of less than 3 hours. Compared to other systems and taking into account what is reported in the literature, the duration time is not increased.

Secondly, you mentioned the American publication on Niastigmine. In my opinion the number of sows used in the USA is fairly limited, while we have hundreds of sows and we could not find any benefit of this product.

S. Baxter *(UK)*

Of course, the two points that you make could be supportive of one another. If it is genetic, if you already have sows that have fairly rapid farrowings, then one would have expected the Niastigmine to have little effect.

A. Hoogerbrugge

However, we have used Niastigmine in special cases where sows have not delivered after four hours but even in these circumstances it was of no use.

B.W. Moss *(UK)*

I have two comments to make in relation to the slide showing bruises and other measurements. I think it is important that whatever measurement we use, whether it is a physiological parameter or a behavioural parameter, that we know the time response change. If we are putting this into practice and we send someone along at 12 days (if I remember correctly) after the period, we would get a complete misinterpretation of the results. I think this is a general statement that would apply to behaviour and physiology - the problem of taking a point measurement which is maybe the way that animal welfare regulations may have to be applied in practice.

The second point is that I would wonder about the misinterpretation of the comment that you made about it being painful on the forelimbs when the pig moves from the standing to the resting position. In my observation, the pig always gets down that way and I think that it could be misinterpreted (although perhaps this is an extreme view) from the point of view that if it is painful then we should prevent pigs from lying down. This is deliberately illogical. I just want to stress that the word painful may lead to other misinterpretations.

S. Baxter

I think that your first point is substantiated by my slide showing that you have to give the time course of injury.

Your second point about what is 'painful': you may recall that when I used the word painful, I hesitated to use it. I did not want it to have an anthropomorphic concept such as has been discussed with play etc. However, what we must remember is that while it is quite natural for the sow to go down that way, it is not natural for the sow to go down like that on a concrete floor. The animal did not evolve this behaviour on a concrete floor. While keeping the word 'painful' in inverted commas, one can hardly doubt that if you take a mass of 150 - 200 kg and distribute that over a piece of the leg which has little muscular tissue across the bone, that in fact the stress distribution will be very high. The only way to overcome this is to develop floor materials where in fact the stress distribution is spread across the knee rather than confined to a narrow point on the knee.

K. Vestergaard *(Denmark)*

I have a comment to make about prolonged farrowing. In my studies we noticed that this prolonged farrowing was due to fixation and only in the sows that were kept loose during pregnancy and single but loose during farrowing and lactation did we have the normal low farrowing time. All the other groups that were fixated in one period or the other, or both, had prolonged farrowing. Since it was not a barren environment it must be the restraint that is important here.

G. Thines *(Belgium)*

When you describe the position of the forelegs of the sow, this is the result of a physical estimation - the objective signs of behaviour, which may be correlated with postulated stress. Did you observe anything typical when the sow was like that?

S. Baxter

We found that when the sow gets into that position she cannot hold the position for long. We timed the sequences through these behaviours. A behavioural action that had a large spread of time showed a point of high stability. The sow could choose to stay there a long time or a short time. When she was on her knees, however, the time scale was invariably very small.

I. Ekesbo *(Sweden)*

Returning to farrowing times, studies made in Sweden on about 650 sows showed that of those that were free in pens, 2.3% had a farrowing time of more than 8 hours. From 1 280 confined sows, 5.4% had a farrowing of more than 8 hours. The only variable factor was confinement.

K. Vestergaard

On the subject of the sow lying down having been on her knees. I noticed in my sows that it took a very long time for the hind part to get down. Do you think that it is a natural adaptation of the domestic sow to be careful not to crush the piglets, or is it because of some leg weakness perhaps?

S. Baxter

There are two things that we have observed. We have found that they take a long time to go down. The position that they reach before going down is a very stable point - they can hold it for some time. When the space is constrained as in a farrowing crate, then the movement that they must make to go down has to be very sudden. In free space this sudden movement never occurs; it becomes a longer and therefore more gradual movement. It is also my belief, although no experiment has been done to prove it, that all these movements are in fact protective ones. Why should the sow develop movements which are not conducive to propagating the species? Along with the lying down movement in the sows in free space in straw pens was a forward movement of the head, brushing the straw to the side as if to make sure that there were no piglets there.

I. Ekesbo

I have often heard in the last ten years of farmers who have had difficulties with MMA (mastitis, metritis, agalactia) syndrome. They have opened the gates and let the sows move around and no more piglets than normal have been lost. This might underline your theories.

S. Baxter

My only comment is that one should not misinterpret what I am saying. I have not said that, having looked at free space, the automatic development for commercial pig production is to go to large pens with plenty of straw. What I have said is that we identify the important functional criteria related to the animal and his behaviour patterns, then we look to see if we can accommodate these in various ways.

GENERAL DISCUSSION

D. Smidt *(FRG)*

I think that the statements made in the previous general discussion on general and methodological aspects of assessing welfare, are also valid for sows and piglets, so in this discussion we should keep to the special areas of concern raised in this session.

Perhaps we can follow again these three areas - reproduction , the pregnant sow, farrowing and suckling . We have learnt that generally more inter-female contacts stimulate reproduction. I do not think that this is necessarily system-dependent, and there is the question of whether you should provide the possibilities for such contacts or not? In Germany, in larger herds, this is provided in special compartments, called in good German 'Eroscentre', and I think this does suitably meet these requirements. However, is this a very urgent welfare problem? I would think it is not as important as some other problems.

In regard to pregnant sows, welfare questions can be associated with the various management systems employed. Undoubtedly, complete confinement in this phase demands special attention, because in combination with complete confinement in the farrowing pen, this means confinement for life. From the point of view of alternatives, there could be more flexibility of management than in, say, farrowing, because the farrowing pens meet a number of functional requirements with respect to farrowing, suckling, and protection of piglets. These pens have been constructed accordingly and are commonly used due to their functional advantages.

The development that has taken place in sow management is not a result of a series of mistakes but has been developed for many good reasons. One reason is the saving of labour requirements in terms of man hours required per sow per year, and this means the labour investment of the farmer and his wife. We should bear this in mind when facing the overall problems.

The main questions that seem to arise from the papers are:

1. Is there enough evidence already to develop final judgements of the acceptability of modern systems? We have not heard contradictory statements on this, but maybe different attitudes. My impression was that the ethological side of these problems was not tackled very intensively, and not enough to cover all the ethological aspects on sows and suckling piglets.

2. What are the perspectives to adapt the systems, if this seems necessary, from the welfare aspects, to animal needs without neglecting the functional necessities of management? Dr. Daelemans gave some examples of such perspectives, and Dr. Baxter defined a huge number of questions involved, which maybe reflect the lack of knowledge in this context. I would like to underline Dr. Ekesbo's statement that we need co-operation, in the field of scientific research on the one hand and on the practical consequences on the other hand. We should keep in mind, in terms of scientific investigations, that animal welfare has not only bio-medical aspects - physiological, ethological and medical aspects - but also technical, economical and ethical aspects, so we need an interdisciplinary approach as we carry out our research.

This is what I meant yesterday when I asked about the possibilities of helping to solve the problems of one discipline by means of the methods of another discipline. I think some misunderstanding arose from this, and I did not mean to interfere with the independence of different disciplines. Once in a while there may be problems in interpreting ethological findings in terms of animal welfare, and maybe physiological correlations might help the interpretations, and vice versa. This is true for ethology and disease problems; veterinarians have always used ethological findings as a basis for their diagnoses. Co-operation is needed to convert the knowledge coming from this research into practical results. I think that the animal welfarists should not be forgotten in this respect, and we should be mutually informed of each other's attitude.

As I said before, I think that we should concentrate on the special problems of sows and suckling piglets, although general points connected with this may also be raised.

J.P. Signoret *(France)*

I think that there is an aspect that we have not discussed. The conditions of the animal under domestication, especially the sow who needs a lot of care, are a result of an interaction between the physical environment and the farmer. Different farmers, working with the same design, and the same environment, can suffer in one case 15% mortality in piglets, and in another case can suffer 3% mortality. In the former case we can say that the system is not in accordance with the welfare of the animal, in the second case we can hypothesise that the same system is providing good welfare for the animals. So this interaction of farmer and environment is an important component of what is the welfare of the animal.

D. Smidt

I agree with you. Would this suggest that training programmes for farmers and herdsmen would contribute in this sense to animal welfare?

J.P. Signoret

Maybe, but one of the best pig herds, for instance, in terms of the number of still-births and mortality among piglets, is using fixed sows. This does not mean that fixation is the best method, but that a herdsman is able to make a given system either acceptable or unacceptable. Training obviously plays a part in this.

D. Smidt

In relation to this, regulations pertaining to animal

welfare can only refer to systems and not to the skills of individual farmers.

J.P. Signoret

Exactly. It is very difficult to formulate regulations that not only take into account the various systems but also the farmer.

I. Ekesbo *(Sweden)*

I think that this is a very interesting sector within the circle that we are discussing and it is relevant when studies are made in single herds. For example, when we have to deal with MMA syndrome which might be a multi-factorial disease, where management may have varying responsibilities. How do you define the possibility for man to compensate for disadvantages in the animal's constant environment? It might be, to some extent, a question of conditioning the animal. I think it important, therefore, that we make very broad investigations on the interactions between farmer and system and link this with legislation.

G. van Putten *(Netherlands)*

Dr. Ekesbo said that we need much more research. If this is not well understood it could be a perfect alibi for doing nothing now. I think that we should be aware that Dr. Ekesbo added that we need more co-operation between veterinarians, ethologists, and technicians before a new housing or management system is allowed to be produced. I think that this is very important for already producing some regulations.

D. Smidt

I agree with you, but this should again be subject to inter-disciplinary co-operation, and I think that right from the beginning we should include technical, and economical aspects as well.

H.C. Adler *(Denmark)*

I have an observation to make about the question of the farmer's skill, how to ensure that the farmer is skilful and has sufficient education to run his farm and manage his animals. In Denmark, we now have legislation according to which a person can only buy a farm if he has a licence, called a green card. To get this card, he must pass courses at agricultural schools and obtain practical experience in the various sectors of agricultural production. In these courses animal behaviour is being introduced - indeed, I have just written a book for this purpose, and this might be a good channel to get more knowledge out onto the farms than in the past.

J.P. Signoret

This might seem somewhat contradictory, but we must also remember that the skill of a farmer or herdsman is not only a matter of learning at school or university, but also due to the 'sense' of animals, knowing whether the animals are well or not. This comes through practice rather than learning. I certainly don't mean that there is no need for education but it is not only education. We must be careful not to produce regulations constraining a good herdsman from developing new systems or practical adaptations in the light of his own experience.

H.C. Adler

The system is used towards practical knowledge.

Doris Buchenauer *(FRG)*

Dr. Signoret, I do not really think that new systems and equipment are created by farmers and herdsmen. In Germany, this equipment is usually developed by people in technical departments, or even industry, who have very little contact with pigs. They should be confronted with their systems in practice.

I would also like to remark to Dr. Ekesbo that 10 years ago, or even more, when we began to discuss these problems, we said then that we had too little information and more investigations must be made. What have we been doing that we still need investigations to come to any conclusions. Of course, we can still know more, but today we have so much information on so many topics that we should be able to make some recommendations for the future, and give information to the people developing the equipment.

I was most impressed by Dr. Baxter's paper which revealed that injuries to the pig were increasing. This is identical to our observations of piglets feeding in intensive systems. We must give some recommendations of what the future systems should look like.

D. Smidt

You are not suggesting that we should produce such recommendations right now?

Doris Buchenauer

I think that we should discuss this. I understood that we would make some conclusions.

D. Smidt

I do not think that 30 minutes gives us enough time for that.

V. Tarrant *(Ireland)*

I think that this meeting may form the basis of future recommendations on pig welfare research in the EEC, and as such we have only given a small amount of attention to the pre-slaughter management of pigs. I think that this is a time in the life of the pig when ownership and management of the animal changes, and there can be quite a high level of suffering, depending upon the particular situation that exists. I would just like to point out that the results of ill-treatment of animals and cruelty to pigs at the pre-slaughter stage has been well recognised by meat quality scientists for a number of years now.

W. Sybesma *(Netherlands)*

I would like to add to Dr. Tarrant's point that castration is also a point that would be of concern to animal welfarists.

J. Connell *(CEC)*

Before you close this session, Mr. Chairman, may I say, on behalf of the Commission, that in 1981 we are planning a meeting to discuss transportation and pre-slaughter procedures. No doubt, the points raised by Dr. Tarrant and Dr. Sybesma will then be discussed in some detail.

FINAL SESSION

Chairman: W. Sybesma

SUMMING UP AND CONCLUSIONS

W. Sybesma *(Netherlands)*

I would like to ask Dr. Huisman to take the floor and present the summing up and the final conclusions.

L.H. Huisman *(Netherlands)*

When Dr. Sybesma and I were asked to organise this symposium he said that he would do the organisation and that the only thing that I would have to do would be the final summary. I did not realise at that time quite what this entailed.

First, I would like to say that we had some excellent contributions and papers. The three Chairmen - Dr. Signoret, Mr. Moss and Dr. Smidt - have given some conclusions and I do not want to repeat everything that they said.

When talking about the welfare of pigs, we must realise, as Dr. Smidt stated, that several disciplines are involved in this - ethology, physiology, veterinarian sciences, genetics, the farm management aspect, farm building and housing research, and so on. Animal welfare is thus multi- rather than uni-disciplined, and these disciplines must co-operate. It was said that to a certain extent animal welfare was not so important, public opinion was really what counted. However, we should never manipulate animals and, therefore, animal welfare should also be seen from the point of view of the animal.

In a number of cases certain provisions for animal welfare may not only be of benefit to the animal but also improve the economic results. In most cases we must assume that animal welfare will cost money. I think that this is acceptable as long as the economic results allow us to pay for it. It is a more difficult question when the economic results cannot afford the price.

What we really need are parameters to measure the effect of the housing systems with which we confront the animals; ethological, environmental, physiological and veterinarian parameters. Doubts have been expressed about the parameters that we are now using, but we must realise that we still need them. At the moment, we do not know enough about the possibilities of ethological and physiological parameters. Probably we know rather more about environmental and veterinarian parameters, the latter especially since they tend to be more exact. Anyway, it is crucial that research in all these areas should continue. No doubt the EEC should help by stimulating additional research. This should be seriously considered. Perhaps a small group could be formed of experts involved in different disciplines to set up a research programme with special priorities for the coming years. A programme of basic research, applied research, and development activities as well. I believe there was a little disappointment that this symposium may not have a follow-up; we all agree that a follow-up is necessary, and this should be encouraged by the Commission.

Mr. Moss has already mentioned Article 3 of the European Convention for the Protection of Animals Kept for Farming Purposes. This says that animals shall be housed and provided with food, water and care, in a manner which - having regard to their species and their degree of development, adaptation and domestication - is appropriate to their physiological and ethological needs in accordance with established experience and scientific knowledge. This is another urgent reason why we need parameters. We must now work on a European level, and we should feel glad to be here together for the first time on a European platform in Brussels, at the EEC headquarters.

There has been a little confusion during these two days. As someone said, collecting information is one thing, interpretation of the data is another thing. Both have to be done, of course.

There has been some discussion on stocking density. It is certainly advisable to decide on minimum space requirements for pigs in different situations, in relation to other relevant aspects. Another important point is the quality of the space. Many papers have mentioned that the presence of straw is beneficial to the animals, in regard to piglets, sows and fattening pigs. Therefore, the development of housing systems to allow varying quantities of straw should be stimulated. I believe that up to now this has had too little attention. There are also possibilities for supplying some straw to fattening pigs on slatted floors. All this development work, already started, should be stimulated. The removal of manure is also very important; and, once again, the proper results can only be achieved with co-operation between research workers of different disciplines.

The criminals are always amongst us. It is too simple to say that tail-biting is only the activity of 'criminals', although they may exist. The disadvantages of tail-biting are obvious, and it would seem that slatted floors contribute to this behaviour. This can be taken into account in the development of new housing systems and as Dr. van Putten suggested, it is not always necessary to wait for all the results of ethological, physiological and other research programmes. In my opinion it would be useful to confront the farm-building and animal housing experts with these facts about tail-biting and urge them to develop new systems that meet more of our present requirements. Maybe this should also be stimulated by the EEC.

Some papers made clear that the pig can adapt to conditions to some extent. Differences were mentioned in the reaction between breeds, and therefore, we should also pay attention to genetic factors.

In most of the countries of the EEC the pigs are kept in what we call intensive systems. In my opinion it would be an illusion to believe that this could be easily changed. However, systems can be developed which are less over-simplified than are many of the present systems.

Finally, let us realise that as far as pigs are concerned, the system of housing is not the only problem. It has been mentioned that for different reasons pigs are given antibiotics, tranquillisers, etc. Therefore, there is the important problem of residues, and this should also be seen as an animal welfare problem because without treatment some systems probably do not work!

Yesterday, one of the participants said to me, "I am confused". Maybe he is still confused, but I hope that after these two days this confusion is on a higher level! In my opinion, it has been a very good Seminar, especially due to the often excellent papers. Thanks are due to the discussion leaders, those who took part in the discussion and all those who worked behind the scenes.

I am at an advantage since there is no discussion after my summary and conclusions. This means that they are my responsibility, and not yours. Let me finish by quoting the Roman statesman, Cicero, who said "I am sorry, I did not have the time to make it shorter".

W. Sybesma

Before we come to the end of this Seminar, I would also like to take the opportunity to thank the discussion leaders, Dr. Signoret, Mr. Moss and Dr. Smidt; also the projectionists and, of course, Janssen Services who have recorded all our discussions. I would like to give the floor now to Dr. Fischbach who is a member of the Standing Committee on Agricultural Research.

<u>W. Fischbach</u> *(Luxembourg)*

I will be very brief, ladies and gentlemen. As a member of the Standing Committee on Agricultural Research, I would like to thank Dr. Sybesma for the excellent organisation of this Seminar. I am convinced that each participant feels, like myself, satisfied by the presentation of such a high standard of papers and the subsequent discussions.

With the exception of Sweden, we are all constrained by the 159 Directives of the Council, which can lead us to ask ourselves whether the requirements for the welfare of pigs will fit with the comparable income of the farmer in the 9 EEC countries, as Dr. Smidt stated. However, I also agree with Dr. Ekesbo, that the gap in income should be paid for by the consumer if we give a better life not only to pigs but to all farm animals.

Finally, I would like to thank the Commission for making this meeting possible.

LIST OF PARTICIPANTS

BELGIUM

Dr. J.M. Bienfait	Faculté de Médecine Vétérinaire 45, rue des Vétérinaires B-1070 Brussels
Dr. J. Connell	Commission of the European Communities DG VI 200 rue de la Loi Brussels
Dr. J. Daelemans	Rijksstation voor Landbouwtechniek Burgemeesters Van Gansberghelaan 115 B-9220 Merelbeke
Dr. W. Goldhorn	Commission of the European Communities 200 rue de la Loi Brussels
Prof. G. Thinès	Centre Albert Michotte Biologie du Comportement B-3041 Pellenberg

DENMARK

Dr. H.C. Adler	Royal Veterinary and Agricultural University Bülowsvej 13 DK-1870 Copenhagen V
Dr. P. Holmegard	Danish Building Research Institute P.O. Box 119 DK-2970 Horsholm
Dr. K. Vestergaard	Royal Veterinary and Agricultural University Bülowsvej 13 DK-1870 Copenhagen V

FEDERAL REPUBLIC OF GERMANY

Prof. Dr. H. Bogner	Bayer. Landesanstalt für Tierzucht Prof. Dürrwaechter-Pl.1 D-8011 Grub/p. Poing
Dr. Doris Buchenauer	Institut für Tierzucht und Tierfütterung der Universität Bonn D-5300 Bonn
Prof. Dr. D. Smidt	Institut für Tierzucht und Tierverhalten D-3057 Neustadt Mariensee
Prof. J. Unshelm	Institut für Tierzucht und Tierverhalten D-2061 Westerau Trenthorst

FRANCE

Dr. R. Dantzer
INRA Laboratoire de Neurobiologie des Comportements
Université de Bordeaux II
146 rue de Léo Saignat
33076 Bordeaux Cedex

Dr. J.Y. Gautier
Laboratoire d'Ethologie
Faculté des Sciences
Université de Rennes Beaulieu
35042 Rennes Cedex

Dr. R. Mauget
Centre d'Etudes Biologiques des Animaux Sauvages
Forêt de Chizé
Villiers en Bois
79360 Beauvoir

Dr. J.P. Signoret
INRA Station de Physiologie de la Reproduction
Centre de Recherches de Tours
37380 Nouzilly

IRELAND

Dr. T. Hanrahan
Pig Husbandry Department
The Agricultural Institute
Moore Park
Fermoy
Co. Cork

Dr. P.V. Tarrant
Dunsinea Research Centre
The Agricultural Institute
Castleknock
Co. Dublin

ITALY

Dr. M. Zanforlin
Università di Padova
Istituto di Psicologia
Piazza Capitaniato 3
35100 Padua

LUXEMBOURG

Dr. V. Fischbach
Administration des Services Techniques de l'Agriculture
16 Route d'Esch
P.O. Box 1904
Luxembourg

Mr. G.J. Breslin
Commission of the European Communities
DG XIII
European Centre
Kirschberg
Luxembourg

THE NETHERLANDS

Ir. R.G. Buré	Institute of Agricultural Engineering P.O. Box 43 6700 AA-Wageningen
Prof. A. Hoogerbrugge	Veterinary Faculty University of Utrecht Yalelaan 17, De Uithof 3584 CL-Utrecht
Ir. L.H. Huisman	Research and Advisory Station for Cattle Husbandry Runderweg 6 8219 PK-Lelystad
Mr. P. Koomans	Institute of Agricultural Engineering P.O. Box 43 6700 AA-Wageningen
Dr. G. van Putten	Research Institute for Animal Husbandry "Schoonoord" P.O. Box 501 3700 AM-Zeist
Dr. W. Sybesma	Research Institute for Animal Husbandry "Schoonoord" P.O. Box 501 3700 AM-Zeist
Dr. W.T. Truyen	Animal Health Centre North Brabant Molenwijkseweg 48 5282 SC-Boxtel

SWEDEN

Prof. I. Ekesbo	Swedish University of Agricultural Sciences Department of Animal Hygiene P.O. Box 345 53200 Skara

UNITED KINGDOM

Dr. S. Baxter	Scottish Farm Buildings Investigation Unit Craibstone Bucksburn Aberdeen AB2 9TR
Mr. S.E.W. Hallam	Janssen Services 33a High Street Chislehurst Kent BR7 5AE
Dr. D.L. Ingram	ARC Institute of Animal Physiology Babraham Cambridge
Dr. B.W. Moss	Department of Agriculture Queens University Belfast

Dr. R. Moss — MAFF
Hook Rise South
Tolworth
Surbiton
Surrey

Mrs. M.J. Robins — Janssen Services
33a High Street
Chislehurst
Kent BR7 5AE

Mr. A. Smith — ADAS Regional Office
Woodthorne
Wolverhampton

Manuscript prepared by:

JANSSEN SERVICES, 33a High Street, Chislehurst, Kent BR7 5AE, UK.